Quantum Physics
Illusion or Reality?

Quantum Physics
Illusion or Reality?

SECOND EDITION

ALASTAIR I. M. RAE

School of Physics and Astronomy
University of Birmingham

CAMBRIDGE
UNIVERSITY PRESS

CAMBRIDGE
UNIVERSITY PRESS

University Printing House, Cambridge CB2 8BS, United Kingdom

One Liberty Plaza, 20th Floor, New York, NY 10006, USA

477 Williamstown Road, Port Melbourne, VIC 3207, Australia

314-321, 3rd Floor, Plot 3, Splendor Forum, Jasola District Centre, New Delhi - 110025, India

103 Penang Road, #05-06/07, Visioncrest Commercial, Singapore 238467

Cambridge University Press is part of the University of Cambridge.

It furthers the University's mission by disseminating knowledge in the pursuit of education, learning and research at the highest international levels of excellence.

www.cambridge.org
Information on this title: www.cambridge.org/9781107604643

First published 1986
Reprinted 1986, 1988, 1989, 1991, 1992
Canto edition 1994
Reprinted 1996, 1998, 2000, 2002, 2004
Second Canto edition 2004
Fifth printing 2009
Reprinted with updates 2009
6th printing 2018

A catalogue record for this publication is available from the British Library

Library of Congress Cataloging in Publication data
Rae, Alastair I. M.
Quantum physics: illusion or reality?
Bibliography
Includes index.
1. Quantum theory.   2. Physics – Philosophy.   I. Title.
QC174.12.R335   1985   530.1´2 85 – 13256

ISBN 978-1-107-60464-3 Paperback

To Ann

I like relativity and quantum theories
Because I don't understand them
And they make me feel as if space shifted
About like a swan that can't settle
Refusing to sit still and be measured
And as if the atom were an impulsive thing
Always changing its mind.

D. H. Lawrence

Time present and time past
Are both perhaps present in time future
And time future contained in time past.

T. S. Eliot

Do you think the things people make fools of
themselves about are any less real and true
than the things they behave sensibly about?

Bernard Shaw

Contents

Preface to the first edition

Quantum physics is the theory that underlies nearly all our current understanding of the physical universe. Since its invention some sixty years ago the scope of quantum theory has expanded to the point where the behaviour of subatomic particles, the properties of the atomic nucleus and the structure and properties of molecules and solids are all successfully described in quantum terms. Yet, ever since its beginning, quantum theory has been haunted by conceptual and philosophical problems which have made it hard to understand and difficult to accept.

As a student of physics some twenty-five years ago, one of the prime fascinations of the subject to me was the great conceptual leap quantum physics required us to make from our conventional ways of thinking about the physical world. As students we puzzled over this, encouraged to some extent by our teachers who were nevertheless more concerned to train us how to apply quantum ideas to the understanding of physical phenomena. At that time it was difficult to find books on the conceptual aspects of the subject – or at least any that discussed the problems in a reasonably accessible way. Some twenty years later when I had the opportunity of teaching quantum mechanics to undergraduate students, I tried to include some references to the conceptual aspects of the subject and, although there was by then a quite extensive literature, much of this was still rather technical and difficult for the non-specialist. With experience I have become convinced that it is possible to explain the conceptual problems of quantum physics without requiring either a thorough understanding of the wide areas of physics to which quantum theory has been applied or a great competence in the mathematical techniques that professionals find so useful. This book is my attempt to achieve this aim.

The first four chapters of the book set out the fundamental ideas of quantum physics and describe the two main conceptual problems: non-locality, which means that different parts of a quantum system appear to influence each other even when they are a long way apart and even although there is no known interaction between them, and the 'measurement problem', which arises from the idea that quantum systems possess properties only when these are measured, although there is apparently nothing outside quantum physics to make the measurement. The later chapters describe the various solutions that have been proposed for these problems. Each of these in some way challenges our conventional view of the physical world and many of their implications are far-reaching and almost incredible. There is still no generally accepted consensus in this area and the final chapter summarises the various points of view and sets out my personal position.

I should like to thank everyone who has helped me in the writing of this book. In particular Simon Capelin, Colin Gough and Chris Isham all read an early draft and offered many useful constructive criticisms. I was greatly stimulated by discussions with the audience of a class I gave under the auspices of the extra-mural department of the University of Birmingham, and I am particularly grateful for their suggestions on how to clarify the discussion of Bell's theorem in Chapter 3. I should also like to offer particular thanks to Judy Astle who typed the manuscript and was patient and helpful with many changes and revisions.

1986

Preface to the second edition

My aims in preparing this second edition have been to simplify and clarify the discussion, wherever this could be done without diluting the content, and to update the text in the light of developments during the last 17 years. The discussion of non-locality and particularly the Bell inequalities in Chapter 3 is an example of both of these. The proof of Bell's theorem has been considerably simplified, without, I believe, damaging its validity, and reference is made to a number of important experiments performed during the last decade of the twentieth century. I am grateful to Lev Vaidman for drawing my attention to the unfairness of some of my criticisms of the 'many worlds' interpretation, and to him and Simon Saunders for their attempts to lead me to an understanding of how the problem of probabilities is addressed in this context. Chapter 6 has been largely rewritten in the light of these, but I am sure that neither of the above will wholeheartedly agree with my conclusions.

Chapter 7 has been revised to include an account of the influential spontaneous-collapse model developed by G. C. Ghiradi, A. Rimini and T. Weber. Significant recent experimental work in this area is also reviewed. There has been considerable progress on the understanding of irreversibility, which is discussed in Chapters 8, 9 and 10. Chapter 9, which emphasised ideas current in the 1980s, has been left largely alone, but the new Chapter 10 deals with developments since then.

This edition has been greatly improved by the input of Chris Timpson, who has read and criticised the manuscript with the eye of a professional philosopher: he should recognise many of his suggested redrafts in the text. I gratefully acknowledge useful discussions with the speakers and other participants at the annual UK conferences on the foundations of physics – in particular Euan Squires whose death in 1996 deprived the foundations-of-physics community of an incisive critical mind and many of us of a good friend. At the editing stage, incisive constructive criticism from Susan Parkinson greatly improved the text. Of course, any remaining errors and mistakes are entirely my responsibility.

1 · Quantum physics

'God', said Albert Einstein, 'does not play dice'. This famous remark by the author of the theory of relativity was not intended as an analysis of the recreational habits of a supreme being but expressed his reaction to the new scientific ideas, developed in the first quarter of the twentieth century, which are now known as quantum physics. Before we can fully appreciate why one of the greatest scientists of modern times should have been led to make such a comment, we must first try to understand the context of scientific and philosophical thought that had become established by the end of the nineteenth century and what it was about the 'new physics' that presented such a radical challenge to this consensus.

What is often thought of as the modern scientific age began in the sixteenth century, when Nicholas Copernicus proposed that the motion of the stars and planets should be described on the assumption that it is the sun, rather than the earth, which is the centre of the solar system. The opposition, not to say persecution, that this idea encountered from much of the establishment of that time is well known, but this was unable to prevent a revolution in thinking whose influence has continued to the present day. From that time on, the accepted test of scientific truth has increasingly been observation and experiment rather than religious or philosophical dogma.

The ideas of Copernicus were developed by Kepler and Galileo and notably, in the late seventeenth century, by Isaac Newton. Newton showed that the motion of the planets resulted directly from two sets of laws: first, the laws of motion, which amount to the statement that the acceleration of a moving body is equal to the force acting on it divided by the body's mass; and, second, the law of gravitation, which asserts that each member of a pair of physical bodies attracts the other by a gravitational force proportional to the product of their masses and inversely proportional to the square of their separation. Moreover, he realised that the same laws applied to the motion of ordinary objects on earth: the apple falling from the tree accelerates because of the force of gravity acting between it and the earth. Newton's work also consolidated the importance of mathematics in understanding physics. The 'laws of nature' were expressed in quantitative form and mathematics was used to deduce the details of the motion of physical systems from these laws. In this way Newton was able not only to show that the motions of the moon and the planets were consequences of his laws but also to explain the pattern of tides and the behaviour of comets.

This objective mathematical approach to natural phenomena was continued in a number of scientific fields. In particular, James Clerk Maxwell in the nineteenth century showed that all that was then known about electricity and magnetism could be deduced from a small number of equations (soon to be known as Maxwell's equations) and that these equations also had solutions in which waves of coupled electric and magnetic fields could propagate through space at the speed of light. This led to the realisation that light itself is just an electromagnetic wave, which differs from other such waves (e.g. radio waves, infrared heat waves, x-rays etc.) only in the magnitudes of its wavelength and frequency. It now seemed that the basic fundamental principles governing the behaviour of the physical universe were known: everything appeared to be subject to Newton's mechanics and Maxwell's electromagnetism.

The philosophical implications of these developments in scientific thought were also becoming understood. It was realised that if everything in the universe was determined by strict physical laws then the future behaviour of any physical system – even in principle the whole universe – could be determined from a knowledge of these laws and of the present state of the system. Of course, exact or even approximate calculations of the future behaviour of complex physical systems were, and still are, quite impossible in practice (consider, for example, the difficulty of forecasting the British weather more than a few days ahead!). But the principle of determinism, in which the future behaviour of the universe is strictly governed by physical laws, certainly seems to be a direct consequence of the way of thinking developed by Newton and his predecessors. It can be summed up in the words of the nineteenth-century French scientist and philosopher Pierre Simon de Laplace: 'We may regard the present state of the universe as the effect of its past and the cause of its future'.

By the end of the nineteenth century, then, although many natural phenomena were not understood in detail, most scientists thought that there were no further fundamental laws of nature to be discovered and that the physical universe was governed by deterministic laws. However, within thirty years a major revolution had occurred that destroyed the basis of both these opinions. These new ideas, which are now known as the quantum theory, originated in the study of electromagnetic radiation, and it is the fundamental changes this theory requires in our conceptual and philosophical thinking which triggered Albert Einstein's comment and which will be the subject of this book. As we shall see, quantum physics leads to the rejection of determinism – certainly of the simple type envisaged by Laplace – so that we have to come to terms with a universe whose present state is not simply 'the effect of its past' and 'the cause of its future'.

Some of the implications of quantum physics, however, are even more radical than this. Traditionally, one of the aims of physics has been to provide an *ontology*, by which is meant a description of physical reality – things as they 'really are'. A classical ontology is based on the concepts of particles, forces and fields interacting under known laws. In contrast, in the standard interpretation of quantum physics it is often impossible to provide such a consistent ontology. For example, quantum theory tells us that the act of measuring or observing an object often profoundly alters its state and that the possible properties of the object may depend on what is actually being measured. As a result, the parameters describing a physical system (e.g. the position, speed etc. of a moving particle) are often described as 'observables', to emphasise the importance of the fact that they gain reality from being measured or 'observed'. So crucial is this that some people have been led to believe that it is the actual human observer's mind that is the only reality – that everything else, including the whole physical universe, is illusion. To avoid this, some have attempted to develop alternative theories with realistic ontologies but which reproduce the results of quantum physics wherever these have been experimentally tested. Others have suggested that quantum physics implies that ours is not the only physical universe and that if we postulate the existence of a myriad of universes with which we have only fleeting interactions, then a form of realism and determinism can be recovered. Others again think that, despite its manifest successes, quantum physics is not the final complete theory of the physical universe and that a further revolution in thought is needed. It is the aim of this book to describe these and other ideas and to explore their implications. Before we can do this, however, we must first find out what quantum physics is, so in this chapter we outline some of the reasons why the quantum theory is needed, describe the main ideas behind it, survey some of its successes and introduce the conceptual problems.

Light waves

Some of the evidence leading to the need for a new way of looking at things came out of a study of the properties of light. However, before we can discuss the new ideas, we must first acquire a more detailed understanding of Maxwell's electromagnetic wave theory of light, to which we referred earlier. Maxwell was able to show that at any point on a light beam there is an electric field and a magnetic field,[1] which are perpendicular both to each other and

---

[1] An electric field exerts a force on a charge that is proportional to the size of the charge. A magnetic field also exerts a force on a charge, but only when it is moving; this force is proportional to both the size of the charge and its speed.

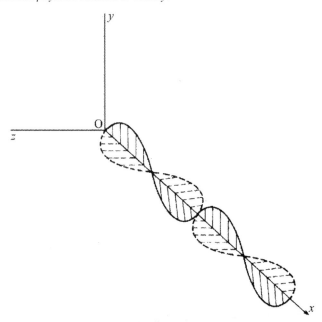

Fig. 1.1 An electromagnetic wave travelling along Ox consists of rapidly oscillating electric and magnetic fields which point parallel to the directions Oy and Oz respectively.

to the direction of the light beam, as illustrated in Figure 1.1. These oscillate many millions of times per second and vary periodically along the beam. The number of oscillations per second in an electromagnetic wave is known as its *frequency* (often denoted by f), while at any point in time the distance between neighbouring peaks is known as the *wavelength* (λ). It follows that the speed of the wave is $c = \lambda f$. The presence of the electric field in an electromagnetic wave could in principle be detected by measuring the electric voltage between two points across the beam. In the case of light such a direct measurement is quite impractical because the oscillation frequency is too large, typically 10^{14} oscillations per second; however, a similar measurement is actually made on radio waves (electromagnetic waves with frequency around 10^6 oscillations per second) every time they are received by an aerial on a radio or TV set.

Direct evidence for the wave nature of light is obtained from the phenomenon known as *interference*. An experiment to demonstrate interference is illustrated in Figure 1.2(a). Light passes through a narrow slit O, after which it encounters a screen containing two slits A and B, and finally reaches a third screen where it is observed. The light reaching a point C on this screen can

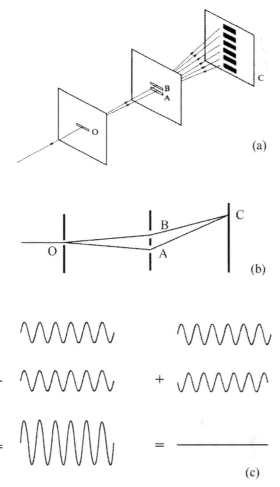

(a)

(b)

(c)

Fig. 1.2 (a) The two-slit interference pattern. (b) Light waves reaching a point C on the screen can have travelled via either of the two slits A and B. The difference in the distances travelled along the two paths is AC − BC. In (c) it is seen that if this path difference equals a whole number of wavelengths then the waves add and reinforce, but if the path difference is an odd number of half wavelengths then the waves cancel. As a result, a series of light and dark bands are observed on the screen, as shown in (a).

have travelled by one of two routes – either by A or by B (Figure 1.2(b)). However, the distances travelled by the light waves following these two paths are not equal, so they do not generally arrive at the point C 'in step' with each other. The difference between the two path distances varies across the pattern on the screen, being zero in the middle. This is illustrated in Figure 1.2(c), from which we see that if the paths differ by a whole number of light wavelengths then the waves reinforce each other, but if the difference is an odd number of half wavelengths then they cancel each other out. Between these extremes the waves partially cancel, so a series of light and dark bands is observed across the screen, as shown in Fig 1.2(a).

The observation of effects such as these 'interference fringes' establishes the wave nature of light. Moreover, measurements on the fringes can be used in a fairly straightforward manner to establish the wavelength of the light used. In this way it has been found that the wavelength of visible light varies as we go through the colours of the rainbow, violet light having the shortest wavelength (about 0.4 millionths of a metre) and red light the longest (about 0.7 millionths of a metre).

Another property of light that will be important shortly is its *intensity*, which, in simple terms, is what we call its brightness. More technically, it is the amount of energy per second carried in the wave. It can also be shown that the intensity is proportional to the square of the amplitude of the wave's electric field, and we will be using this result below.

Photons

One of the first experiments to show that all was not well with 'classical' nineteenth-century physics was the *photoelectric effect*. In this, light is directed on to a piece of metal in a vacuum and as a result subatomic charged particles known as electrons are knocked out of the metal and can be detected by applying a voltage between it and a collector plate. The surprising result of such investigations is that the energy of the individual emitted electrons does not depend on the brightness of the light, but only on its frequency or wavelength. We mentioned above that the intensity or brightness of light is related to the amount of energy it carries. This energy is transferred to the electrons, so the brighter the light, the more energy the body of escaping electrons acquires. We can imagine three ways in which this might happen: each electron must acquire more energy, or there must be more electrons emitted or both things happen. In fact, the second possibility is the one that occurs: for light of a given wavelength, the *number* of electrons emitted per second increases with the light intensity, but the amount of energy acquired by each individual electron is unchanged. However strong or weak the light, the energy given to each escaping electron equals hf, where f is the frequency of the light wave and *h* is a universal constant of quantum physics known as Planck's constant.

The fact that the electrons seem to be acquiring energy in discrete bits and that this can only be coming from the light beam led Albert Einstein (the same scientist who developed the theory of relativity) to conclude that the energy in a light beam is carried in packets, sometimes known as 'quanta' or 'photons'. The value of *hf* is very small and so, for light of normal intensity, the number of packets arriving per second is so large that the properties of such a light beam are indistinguishable from those expected from a continuous wave. For example, about 10^{12} (a million million) photons per second pass through an area the size of a full stop on this page in a typically lit room. It is only the very particular circumstances of experiments such as the photoelectric effect that allow the photon nature of light to be observed.

We can get further insight into the properties of photons by considering experiments where the incident light is very weak. If the light were simply a wave, we would not expect any electrons to be emitted until wave energy amounting to at least *hf* had arrived at one of the atoms. However, we actually find that some electrons are detected immediately after the light is switched on, and well before enough energy could have been supplied by the light wave. The conclusion to be drawn from this is that the photon energy must be carried in a small volume so that, even if the average rate of arrival of photons is low, there will be a reasonable chance that one of them will release its energy to an electron early in the process. In this sense at least, the photon behaves like a small *particle*. Further work confirmed this: for example, photons were seen to bounce off electrons and other objects, conserving energy and momentum and generally behaving just like particles rather than waves.

We therefore have two models to describe the nature of light, depending on the way we observe it: if we perform an interference experiment then light behaves as a wave, but if we examine the photoelectric effect then light behaves like a stream of particles. Is it possible to reconcile these two models?

One suggestion for a possible reconciliation is that we were mistaken ever to think of light as a wave. Perhaps we should always have thought of it as a stream of particles with rather unusual properties that give rise to interference patterns, so that we were simply wrong ever to describe it using a continuous-wave model. This would mean that the photons passing through the apparatus shown in Figure 1.2 would somehow bump into each other, or interact in some way, so as to guide most of the photons into the light bands of the pattern and very few into the dark areas. This suggestion, although elaborate, is not ruled out by most interference experiments because there is usually a large number of photons passing through the apparatus at any one

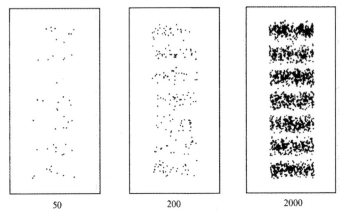

50   200   2000

Fig. 1.3 The three panels show a computer reconstruction of the appearance of a two-slit interference pattern after 50, 200 and 2000 photons respectively have arrived at the screen. The pattern appears clear only after a large number of photons have been recorded even though these have passed through the apparatus one at a time.

time and interactions are always conceivable. If however we were to perform the experiment with very weak light, so that at any time there is only one photon in the region between the first slit and the screen, interactions between photons would be impossible and we might then expect the interference pattern to disappear. Such an experiment is a little difficult, but perfectly possible. The final screen must be replaced by a photographic plate or film and the apparatus must be carefully shielded from stray light; but if we do this and wait until a large number of photons has passed through one at a time, the interference pattern recorded on the photographic plate is just the same as it was on the screen in the earlier experiment!

We could go a little further and repeat the experiment several times using different lengths of exposure. We would then get results like those illustrated in Figure 1.3, from which we see that the photon nature of light is confirmed by the appearance of individual spots on the photographic film. At very short exposures these seem to be scattered more or less at random, but the interference pattern becomes clearer as more and more arrive. We are therefore forced to the conclusion that interference does not result from interactions between photons; rather, *each photon* must undergo interference at the slits A and B. Indeed, the fact that the interference pattern created after a long exposure to weak light is identical to one produced by the same number of photons arriving more or less together in a strong light beam implies that photons may not interact with each other at all.

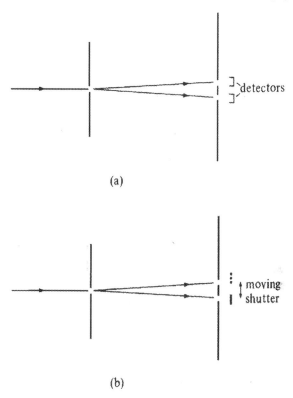

Fig. 1.4 If we place photon detectors behind the two slits of an interference apparatus, as in (a), each photon is always recorded as passing
through one slit or the other and never through both simultaneously. If,
as in (b), a shutter is placed behind the slits and oscillated up and down
in such a way that both slits are never open simultaneously, the two-slit
interference pattern is destroyed.

If interference does not result from interaction between photons, could it
be that each individual photon somehow splits in two as it passes through the
double slit? We could test for this if we put a photographic film or some kind
of photon detector immediately behind the two slits instead of some distance
away. In this way we could tell through which slit the photon passes, or
whether it splits in two on its way through (see Figure 1.4). If we do this,
however, we always find that the photon has passed through one slit or the
other and we never find any evidence that the photon splits. Another test of
this point is illustrated in Figure 1.4(b): if a shutter is placed behind the two
slits and oscillated up and down so that only one of the two slits is open at any

one time, the interference pattern is destroyed. The same thing happens when any experiment is performed that detects, however subtly, through which slit the photon passes. It seems that light passes through one slit or the other in the form of photons if we set up an experiment to detect through which slit the photon passes, but passes through both slits in the form of a wave if we perform an interference experiment.

The fact that processes like two-slit interference require light to exhibit both particle and wave properties is known as *wave–particle duality*. It illustrates a general property of quantum physics, which is that the nature of the model required to describe a system depends on the nature of the apparatus with which it is interacting. Light has the property of a wave when passing through a pair of slits but has to be considered as a stream of photons when it strikes a detector or a photographic film. This dependence of the properties of a quantum system on the nature of the observation being made on it underlies the conceptual and philosophical problems that it is the purpose of this book to discuss. We shall begin this discussion in a more serious way in the next chapter, but we devote the rest of this chapter to a discussion of some further implications of the quantum theory and to an outline of some of its outstanding successes in explaining the behaviour of physical systems.

The Heisenberg uncertainty principle

One of the consequences of wave–particle duality is that it sets limits on the amount of information that can be obtained about a quantum system at any one time. Thus we can *either* choose to measure the wave properties of light by allowing it to pass through a double slit without detecting through which slit the photon passes *or* we can observe the photons as they pass through the slits. We can never do both these things at once. Werner Heisenberg, one of the physicists who were instrumental in the early development of quantum physics, realised that this type of measurement and its limitations could be described in a rather different way. We can think of identifying which slit a photon went through as essentially a measurement of the position of the photon as it passes through the slits, while the observation of interference is akin to a measurement of its momentum. It follows from wave–particle duality that it is impossible to make simultaneous precise measurements of the position and momentum of a quantum object such as a photon.

The application of Heisenberg's ideas to the two-slit experiment is actually rather subtle, and a more straightforward example is the behaviour of light passing through a single slit of finite width. If this is analysed using the wave model of light, then we find, as shown in Figure 1.5, that the slit spreads the light out into a 'diffraction pattern'. We also find that if we make the slit in the screen narrower, the diffraction pattern on the screen becomes

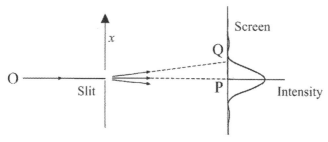

Fig. 1.5 Light passing through a single slit is diffracted to form a diffraction pattern whose intensity varies in the manner illustrated by the graph on the right. The narrower the slit, the broader is the diffraction pattern. As explained in the text, this result leads to limits on the possible accuracy of simultaneous measurements of the position and momentum of the photons that are in accordance with Heisenberg's uncertainty principle.

broader. We can perform this experiment with very weak light so as to study the behaviour of individual photons just as in the two-slit case – cf. Figure 1.3. We again find that individual photons arrive at more or less random points on the screen, but the diffraction pattern is established as more and more photons accumulate.

The uncertainty principle relates to what we are able to predict about the properties of a photon passing through this apparatus. We know that it must pass through the slit, but we do not know where. There is therefore an *uncertainty* in the particle position and the size of this uncertainty is Δx, where Δx is the width of the slit. After the particle leaves the slit it will arrive somewhere on the screen, but again we do not know beforehand where that will be. We can say that if the particle strikes the screen in the centre, its motion between the slit and the screen must have been along the horizontal line OP. However, if it arrives at a point away from the centre of the pattern – say the point Q – its velocity must have been at an angle to the horizontal. In this case, the particle's velocity, and hence its momentum, must have had a component in the x direction. Since all we know is that the photon will arrive somewhere in the diffraction pattern, the width of this diffraction pattern is a measure of the uncertainty in our prediction of this component of momentum.

It follows that, if we make the slit smaller, in order to reduce the uncertainty in position, we will inevitably increase the spread of the diffraction pattern and, correspondingly, the momentum uncertainty in the x direction. It turns out that, if we multiply these two uncertainties together, the result is the same for all slit sizes, and if we perform a more detailed calculation, this quantity is found to be approximately equal to the fundamental quantum constant (i.e. Planck's constant, *h*, mentioned above). Heisenberg was able to show that quantum theory requires that all such predictions of position and momentum are subject to similar limitations, the products of their uncertainties never being less than *h* divided by 4π. He expressed this in his famous *uncertainty principle*, in which the uncertainty (Δ*x*) in position is related to that in momentum (Δ*p*) by the relation

$$\Delta x \, \Delta p > \frac{h}{4\pi}$$

Clearly our analysis of the single-slit diffraction case is consistent with this.

The implications of the uncertainty principle on the way we think about scientific measurement are profound. It had long been realised that there are practical limitations to the accuracy of any measurement, but before quantum physics there was no reason in principle why we should not be able to attain any desired accuracy by improving our experimental techniques. Although the uncertainty principle relates to our ability to *predict* the results of subsequent measurements, in practice it also puts a fundamental limit on the precision of any simultaneous measurement of two physical quantities, such as the position and momentum of a photon. After this idea was put forward, there were a number of attempts to suggest experiments that might be capable of making measurements more precisely than the uncertainty principle allows, but in every case careful analysis showed that this was impossible. As we shall see in later chapters, in the standard interpretation of quantum physics the whole concept of attributing a definite position to a particle of known momentum is invalid and meaningless. The uncertainty principle is just one of the many strange and revolutionary consequences of quantum physics that have led to the conceptual and philosophical ideas that are the subject of this book.

Atoms and matter waves

Just as the wave model of light was well established in classical physics, there was little doubt by the beginning of the twentieth century that matter was made up of a large number of very small particles or *atoms*. Dalton's atomic theory had been remarkably successful in explaining chemical processes, and the phenomenon of Brownian motion (in which smoke particles suspended in air are observed to undergo irregular fluctuations) had been explained as a consequence of the random impacts of air molecules. The study of the properties of electrical discharge tubes (the forerunners of the cathode-ray tube found in television sets) led J. J. Thompson to conclude that electrically charged particles, soon afterwards called electrons, are emitted when a metal wire is heated to a high temperature in a vacuum. Very early in the twentieth century, Ernest Rutherford showed that the atom possesses a very small positively charged nucleus in which nearly all the atomic mass is concentrated, and he then suggested that the atom consists of this nucleus surrounded by electrons. At this point a problem arose. Every attempt to describe the structure of the atom in more detail using classical physics failed. An obvious model was one in which the electrons orbit the nucleus as a planet orbits the sun. However, Maxwell's electromagnetic theory requires that such an orbiting charge should radiate energy in the form of electromagnetic waves and, as this energy could come only from the electrons' motion, these would soon slow up and fall into the nucleus.

Just before World War I, the Danish physicist Niels Bohr, of whom we shall be hearing much more in due course, devised a model of the hydrogen atom (which contains only a single electron) in which such electron orbits were assumed to be stable under certain conditions, and this model had considerable success. However, it failed to account for the properties of atoms containing more than one electron and there was no rationale for the rules determining the stability of the orbits. A decade later, in the early 1920s, the French physicist Louis de Broglie put forward a radical hypothesis. If light waves sometimes behave like particles, could it be that particles, such as electrons and nuclei, sometimes exhibit wave properties? To test such an apparently outrageous idea we might think of passing a beam of electrons through a two-slit apparatus of the type used to demonstrate interference between light waves (Figure 1.2). At the time this was not practicable, because the wavelengths predicted by de Broglie for such electron beams were so short that the interference fringes would be too close together to be observed. However, de Broglie's idea was tested by a very similar experiment in which electrons were scattered from a nickel crystal. An intensity pattern was observed which could be explained by assuming that interference had occurred between the electron waves scattered by different planes of atoms in the crystal and that the electron beam had indeed behaved like a wave in this situation. Much more recently it has been possible to generate electron beams of longer wavelength so as to demonstrate two-slit interference directly. Similar experiments performed with other particles, such as neutrons, atoms and molecules, have confirmed that these also have wave properties.

The matter-wave hypothesis was also confirmed indirectly, though possibly more dramatically, by its ability to explain the electronic structure of atoms. A proper understanding of this point requires mathematical analysis well beyond the scope of this book, but the essence of the argument is that when matter waves are confined within a region of space only particular wavelengths are allowed. On an everyday analogy, a violin string of given length and tension can emit only particular notes, and, indeed, similar principles govern the operation of most musical instruments.

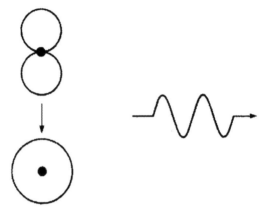

Fig. 1.6 Two of the possible stable patterns adopted for electron waves in atoms are indicated on the left. If the atom makes a transition from the upper (higher-energy) to the lower state, a light photon of definite wavelength is emitted.

It is found that when the matter-wave hypothesis is combined with the fact that the negative electrons are attracted to the positive nucleus by an inverse square law of force, an equation results whose solutions determine the form of the electron waves in this situation. This equation (known as the Schrödinger equation after Erwin Schrödinger, who devised it) has solutions for only particular 'quantised' values of the electron energy. It follows that an electron in an atom cannot have an energy lower than the lowest of these allowed values (known as the 'ground state energy') so the problem of the electrons' spiralling into the nucleus is avoided. Moreover, if an atom is 'excited' into an allowed state whose energy is higher than that of the ground state, it will jump back to the ground state while emitting a photon whose energy is equal to the difference between the energies of the two states (Figure 1.6). We saw earlier that the energy of a photon is closely related to the wavelength of the associated light wave, so it follows that light is emitted by atoms at particular wavelengths only. It had been known for some time that light emitted from atoms (in gas-discharge tubes for example) had this property, and it is a major triumph of the Schrödinger equation that not only can this be explained but also the magnitudes of the allowed wavelengths can be calculated, and they are found to be in excellent agreement with experiment.

Beyond the atom

The success of the matter-wave model did not stop at the atom. Similar ideas were applied to the structure of the nucleus itself, which is known to contain an assemblage of positively charged particles, called protons, along with an approximately equal number of uncharged 'neutrons'; collectively protons and neutrons are known as nucleons. There is a strong attractive force between all nucleons, known as the 'strong interaction', which exists in addition to the electrostatic repulsion between protons. Its form is quite complex and indeed is not known precisely, so the calculations are considerably more difficult than in the atomic case. The results, however, are just as good: the calculated properties of atomic nuclei are found to be in excellent agreement with experiment.

Nowadays even 'fundamental' particles such as the proton and neutron (but not the electron) are known to have a structure and to be composed of even more fundamental objects known as 'quarks'. This structure has also been successfully analysed by quantum physics in a manner similar to that for the nucleus and the atom, showing that the quarks also possess wave properties. But modern particle physics has extended quantum ideas even beyond this point. At high enough energies a photon can be converted into a negatively charged electron along with an otherwise identical, but positively charged, particle known as a positron, and electron–positron pairs can recombine into photons. Moreover, exotic particles can be created in high-energy processes, many of which spontaneously decay after a small fraction of a second into more familiar stable entities such as electrons or quarks. All such processes can be understood by an extension of quantum ideas into a form known as quantum field theory. An essential feature of this theory is that some phenomena are best described as a superposition of a number of fundamental processes, analogously to the superposition of the waves passing through the two slits of an interference apparatus.

Condensed matter

The successes of quantum physics are not confined to atomic or subatomic phenomena. Soon after the establishment of the matter-wave hypothesis, it became apparent that it could also be used to explain chemical bonding. For example, in the case of a molecule consisting of two hydrogen atoms, the electron waves surround both nuclei and draw them together, with a force that is balanced by the mutual electrical repulsion of the positive nuclei, to form the hydrogen molecule. These ideas can be developed into calculations of molecular properties, such as the equilibrium nuclear separation, which agree precisely with experiment.

The application of similar principles to the structure of condensed matter, particularly crystalline solids, has been just as successful. The atoms in a crystal are arranged on a regular lattice, and one of the properties of such a lattice is that it scatters waves passing through it, if these have wavelengths that are related to the distances between the planes of atoms in the crystal. Otherwise the waves pass through the lattice largely undisturbed. We mentioned an example of this earlier, when we cited the observation of diffraction when a beam of electrons strikes a crystal as evidence for De Broglie's proposed matter waves. It turns out that in a metal some of the electrons (typically one or two per atom) are not attached to the atoms but are free to move through the whole crystal. Moreover, the wavelengths of the associated electron waves are too long for them to be diffracted by the crystal lattice. As a result, the waves move through the crystal unhindered, resulting in an electric current flow with little resistance. In contrast, the wavelengths in insulators are shorter and the electron flow is blocked completely by the occurrence of diffraction. In 'semiconductors' such as silicon, only a small fraction of the electrons are free to move and this leads to the special properties evinced by the silicon chip with all its ramifications. Moreover, the exotic properties of materials at very low temperatures, where liquid helium has zero viscosity and some metals become superconductors completely devoid of electrical resistance, can be shown to be manifestations of quantum behaviour, and we will return to this briefly in Chapter 7.

The last three sections of this chapter have only touched on some of the manifest successes quantum physics has achieved over the last half-century. Wherever it has been possible to perform a quantum calculation of a physical quantity it has been in excellent agreement with the results of experiment. However, the purpose of this book is not to survey this achievement in detail, but rather to explore the fundamental features of the quantum approach and to explain their revolutionary implications for our conceptual and philosophical understanding of the physical world. To achieve this we need a more detailed understanding of quantum ideas than we have obtained so far, and we begin this task in the next chapter.

2 · Which way are the photons pointing?

The previous chapter surveyed part of the rich variety of physical phenomena that can be understood using the ideas of quantum physics. Now that we are beginning the task of looking more deeply into the subject we shall find it useful to concentrate on examples that are comparatively simple to understand but which still illustrate the fundamental principles and highlight the basic conceptual problems. Some years ago most writers discussing such topics would probably have turned to the example of a 'particle' passing through a two-slit apparatus (as in Figure 1.2), whose wave properties are revealed in the interference pattern. Much of the discussion would have been in terms of wave–particle duality and the problems involved in position and momentum measurements, as in the discussion of the uncertainty principle in the last chapter. However, there are essentially an infinite number of places where the particle can be and an infinite number of possible values of its momentum, and this complicates the discussion considerably. We can illustrate all the points of principle we want to discuss by considering situations where a measurement has only a small number of possible outcomes. One such quantity relating to the physics of light beams and photons is known as *polarisation*. In the next section we discuss it in the context of the classical wave theory of light, and the rest of the chapter extends the concept to situations where the photon nature of light is important.

The polarisation of light

Imagine that a beam of light is coming towards us and that we think of it as an electromagnetic wave. As we saw in Chapter 1 (Figure 1.1) this means that at any point in space along the wave there is an electric field that is vibrating many times per second. At any moment in time, this electric field must be pointing in some direction, and it turns out that Maxwell's equations require the direction of vibration always to be at right angles to the direction of travel of the light. So if the light is coming towards us the electric field may point to the left or the right or up or down or in some direction in between, but not towards or away from us (Figure 2.1). In many cases the plane containing the electric field direction changes rapidly from time to time, but it is possible to create light in which this plane remains constant. Such light is said to be *plane polarised* or sometimes just *polarised*. The plane containing the

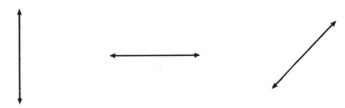

Fig. 2.1 In a light wave coming towards us the electric field may oscillate vertically, horizontally or at some angle in between, but the oscillation is always perpendicular to the direction of travel of the light beam.

electric field vectors is known as the *plane of polarisation* and the direction in which the electric field points is known as the *polarisation direction*.

The idea of polarisation may be more familiar to some readers in the context of radio or TV reception. To get a good signal into a receiver it is necessary to align the aerial dipole along the polarisation direction (usually either horizontal or vertical) of the radio waves. This ensures that the electric field will drive a current along the aerial wire and hence into the set.

Polarised light can be produced in a number of ways. For example, light from most lasers is polarised as a result of internal processes in the laser, and a polarised beam can be conveniently produced from any beam of light using a substance known as 'Polaroid'. This substance is actually a thin film of nitrocellulose packed with extremely small crystals, but the construction and operational details are not relevant to our discussion. What is important is that if ordinary unpolarised light shines on one side of a piece of Polaroid, the light emerging from the other side is polarised and has an intensity about half that of the incident light (Figure 2.2). The polarisation direction of the light coming out of a Polaroid is always along a particular direction in the Polaroid sheet known as the Polaroid axis. Readers may well be familiar with the use of Polaroid in sunglasses: the bright daylight is largely unpolarised, so only about half of it gets through to the eye.

Polaroid can also be used to find the polarisation direction of light that is already polarised: we just rotate the Polaroid about the direction of the light beam until the emerging light is a maximum and nearly as strong as the light that went in. A very important point to note is that the Polaroid axis does not have to be exactly lined up with the polarisation direction before any light comes through at all. The light is stopped completely only if the two are at right angles, and the transmitted fraction increases gradually and smoothly as the Polaroid is rotated. In slightly more technical language, we say that the Polaroid allows through the *component* of the light that is polarised in the

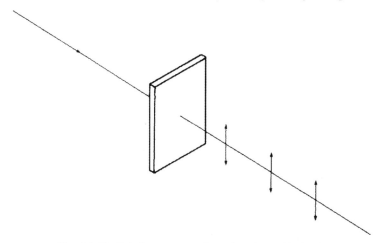

Fig. 2.2 If a light beam passes through a piece of Polaroid the electric vector of the emitted light is always parallel to a particular direction (vertical in the case shown) known as the Polaroid axis.

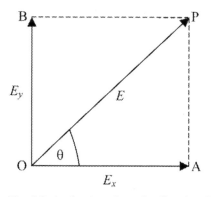

Fig. 2.3 A vibration along the direction OP can be thought of as a combination of vibrations along OA and OB. If light with electric field amplitude OP (intensity OP^2) is passed through a Polaroid with its axis along OA, the amplitude of the transmitted light will be equal to OA and its intensity will be OA^2. Note also that OA = OP cos θ, where θ is the angle POA.

direction of the Polaroid axis. This is illustrated in Figure 2.3, which shows how an electric field in a general direction (OP) can be thought of as the addition of two components (OA and OB) at right angles to each other. If the Polaroid axis points along OA, say, the component in this direction will pass through while the component along OB will be absorbed. Hence, the

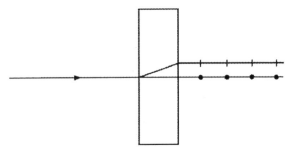

Fig. 2.4 Light passing through a crystal of calcite is divided into two components whose polarisations are respectively parallel (upper beam) and perpendicular (lower beam) to a particular direction in the calcite crystal.

direction and magnitude of the electric field of the emergent light are the same as those of the component of the incident light parallel to the Polaroid axis. The particular cases where the angle POA $= 0°$ and $90°$ clearly correspond to transmission of all and none of the light respectively. In general, the intensity of the emergent light is equal to $I\cos^2\theta$, where $I = E^2$ is the brightness or intensity (see Chapter 1) of the incident intensity and θ is the angle POA.

A slightly different kind of device used for generating and analysing the polarisation of light is a single crystal of the mineral calcite. The details of operation need not concern us, but such a crystal is able to separate the light into two components with perpendicular polarisation (Figure 2.4). Unlike Polaroid, which absorbs the component perpendicular to the Polaroid direction, the calcite crystal allows all the light through, but the two components emerge along different paths. Because the total light emerging in the two beams is equal to that entering the calcite crystal and because light intensity is proportional to the square of the electric field, it follows that the incident intensity E^2 equals the sum of the two transmitted intensities E_x^2 and E_y^2 by Pythagoras' theorem (see Figure 2.3).

Consideration of the analysis of a light beam into two polarisation components by a device such as a calcite crystal will play a central role in much of the discussion in later chapters. However, as the details of how this is achieved are not of importance, from now on we shall illustrate the process simply by drawing a square box with one incident and two emergent light beams, as in Figure 2.5. The label 'H/V' on the box shows that it is oriented in such a way that the emergent beams are polarised in the horizontal and vertical directions; we shall also consider other orientations such as $\pm45°$, which means that the emergent beams are polarised at $+45°$ and $-45°$ to the horizontal respectively. In discussing polarisation, we will represent a

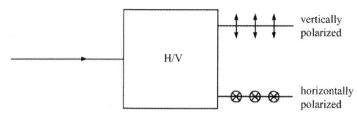

Fig. 2.5 In discussing polarisation, we represent a polariser (such as that illustrated in Figure 2.4) by a box with a label indicating the directions of the polarisation axes. In the example shown, the box is marked H/V because it resolves the incident light into horizontally and vertically polarised components.

polariser (such as that illustrated in Figure 2.4) by a box with a legend indicating the directions of the polarisation axes. In the example shown in Figure 2.5, the box resolves the incident light into components polarised in the vertical and horizontal directions.

The polarisation of photons

We saw in the last chapter that classical wave theory is unable to provide a complete description of all the properties of light. In particular, when a device based on the photoelectric effect detects light, the light behaves as if it consisted of a stream of particles, known as photons. The photon nature of light is particularly noticeable if the overall intensity is very low, so that the arrival of photons at the detector is indicated by occasional clicks: at higher intensities the clicks run into each other and the behaviour is the same as would be expected from a continuous wave. At first sight it might seem that polarisation is very much a wave concept and might not be applicable to individual photons. However, consider the experiment illustrated in Figure 2.5. If a very weak beam of unpolarised light is incident on a polariser and the two output beams are directed onto detectors capable of counting individual photons, the photons must emerge in one or other of the two channels – if only because they have nowhere else to go! We can therefore label the individual photons, calling those emerging in the H channel 'h' and those in the V channel 'v'. Performing further measurements using additional H/V polarisers, as shown in Figure 2.6, tests whether these labels reflect any physical property of the photons themselves. It turns out that the h and v photons invariably emerge in the H and V channels respectively of the second polarisers. We therefore have what is known as an 'operational' definition of photon polarisation. That is, although we may not be able to say what it is, we can define photon polarisation in terms of the operations that

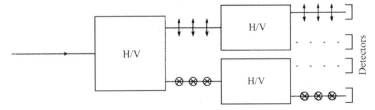

Fig. 2.6 Polarisation is a property that can be attributed to photons because every photon that emerges from the first polariser as vertically or horizontally polarised passes through the corresponding channel of the subsequent H/V polariser.

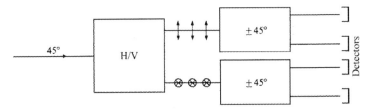

Fig. 2.7 If 45° polarised photons are incident on an H/V polariser they emerge as either horizontally or vertically polarised. They then pass at random through the two channels of the ±45° polarisers, after which they are detected, so they appear to have lost any 'memory' of their original polarisation. We conclude that polarisation measurements in general change the polarisation state of the measured photons.

have to be performed in order to measure it. Thus a horizontally polarised photon, for example, is one that has emerged through the H channel of an H/V calcite crystal or through a Polaroid whose axis is in the horizontal direction.

We can now see how the measurement of photon polarisation illustrates some of the general features of quantum measurement. We saw in the previous chapter how a measurement of some physical property, such as a particle's position, inevitably affected the system in such a way that the results of measuring some other property, such as the particle's momentum, become unpredictable. The same principle arises when we consider the measurement of the H/V polarisation of a photon that was previously known to have 45° polarisation. Consider the set-up shown in Figure 2.7. A beam of photons, all polarised at 45° to the horizontal (obtained perhaps by passing an unpolarised light beam through an appropriately oriented Polaroid) is incident on a polariser oriented to measure H/V polarisation. Half the photons emerge in the horizontal channel and the other half in the vertical channel. If we now pass either beam through another polariser oriented to perform $\pm 45°$ measurements, we find that the original $45°$ polarisation has been destroyed by the H/V measurement: half the photons in each of the H/V beams emerge through each $\pm45°$ channel. Measuring the H/V polarisation has therefore altered the state of the photons so that their $\pm45°$ polarisations are now unknown until these are measured again. Similarly, a $45°$ measurement would alter the state of a photon whose H/V polarisation is known. The same conclusion can be drawn about successive measurements of any two polarisation components. Referring to Figure 2.3 and the earlier discussion relating to waves, we consider the case of a beam of (say) h photons approaching a polariser that is oriented to measure polarisation parallel and perpendicular to a direction making an angle θ with the horizontal. In the wave case, we saw that the intensities of the beams in the parallel and perpendicular channels are respectively E_x^2 and E_y^2 and therefore proportional to $\cos^2\theta$ and $\sin^2\theta$ respectively. It follows that if we were to perform this experiment on a large number (say N) of photons the number emerging in each channel would be $N\cos^2\theta$ and $N\sin^2\theta$. The corresponding probabilities of emergence for one photon are $\cos^2\theta$ and $\sin^2\theta$. However, until we actually carry out the experiment this is all we can know, as the actual result for any particular photon is in general random and unpredictable. (The only exception is the special case where the second apparatus is also in the H/V orientation: then a second measurement can be made without disturbing a previously known polarisation state of the photons).

Let us make a more careful comparison of the behaviour of light waves and photons in this context. An electromagnetic wave polarised at, say, $45°$ to the horizontal can be considered as being made up of horizontal and vertical components as in Figure 2.3 and the effect of an H/V polariser is to split the wave into these two parts, which then proceed as separate beams. The analogous statement for an individual photon is that a $+45°$ photon is in a state that is a *superposition* of a state corresponding to an h photon and one corresponding to a v photon and, when we pass the photon through an H/V polariser, we detect it as either h or v and say that the superposition has 'collapsed' into one of these states. It is important to realise that describing the photon's state as a superposition does not of course change this state before it is measured. It is still a $+45°$ photon, but it can be convenient for us to represent the state as a superposition, in the same way that it can be convenient to resolve a wave into two components that add together to produce the original polarisation. Moreover, the only reason for referring the superposition to the H and V directions in particular is that we know we are going to consider an H/V measurement. It would be just as valid to express the wave as a sum of its components along any pair of mutually perpendicular directions and we can choose whichever are most appropriate for the situation we are considering. We can also change our reference axes as the experiment evolves; for example, in Figure 2.7 we would express the original 45° state as a superposition of h and v that collapses into h or v, but then we would express h and v as superpositions of +45° and −45° before considering the second set of measurements.

The concept of superposition will play an important role in our later discussions, so it is important to understand what it means. We emphasise again that, *before* the measurement, a photon polarised at 45° to the horizontal is neither horizontally nor vertically polarised: it is in a superposition of both states. The question 'is it horizontally or vertically polarised?' is meaningless and unanswerable, in the same way as it would be meaningless to ask whether a 45° wave was horizontally or vertically polarised. *After* the measurement, however, the state of the particle collapses into *either h or v*, so the above question is now meaningful and can be answered by noting through which channel photon has come out. Once this has taken place, it becomes equally meaningless to ascribe +45° or −45° polarisation to the h or the v photon, although we can describe either as being in a ±45° superposition.

We should note that the superposition concept can be generalised to the measurement of any quantum property; for example, a particle in a definite momentum state can be described as being in a superposition of all possible position states and vice versa. We should also note that this analysis, although orthodox, is not universally accepted and we shall be introducing some alternative interpretations later in this chapter.

We see now one of the ways in which quantum physics profoundly changes our traditional way of thinking about the way the physical universe works and, in particular, about the question of determinism. When a 45° polarised photon approaches an H/V polariser, we know that it will come out in one channel or the other, but there is no way of telling in advance which this will be. If it is meaningless to ascribe an H/V polarisation to the photon before it enters this apparatus, we can conclude similarly that the outcome of the measurement is determined purely by chance. The result of the measurement – i.e. the channel through which the photon will emerge – is completely unpredictable. We know that if there are many photons then on average half of them will appear in each channel, but if we concentrate on one photon only then its behaviour is completely random. Indeed (although we must be careful about using such language) we could say that even the photon does not know through which channel it is going to emerge! We should note that this unpredictability is a result of the presence of photons, and does not arise in the continuous wave model. An H/V polariser allows half of the incoming 45° polarised wave to pass along each channel and we don't need to worry about which half; it is only because an

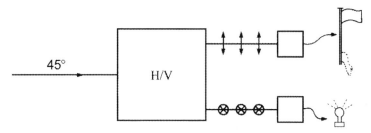

Fig. 2.8 The channel through which a photon will emerge is determined purely by chance. If an outgoing vertically polarised photon triggers a device to raise a flag while an outgoing horizontal photon causes a light to be switched on, then which of these two events occurs for a given incident photon is completely unpredictable and undetermined.

individual photon has to pass through either one channel or the other that the indeterminacy arises.

However, although the indeterminacy results from the behaviour of the photon, it is important to realise that it need not stop there. If we consider a single 45° photon passing through an H/V apparatus, the fact that it emerges at random through one or the other channel means that we cannot predict which detector will record an event. But the operation of a detector is not a microscopic event; it operates on the large scale of the laboratory or of everyday events. Indeed there is no reason why the operation of the detectors should not be coupled to some more dramatic happening, such as the flashing of a light or the raising of a flag (Figure 2.8). If we consider a single photon entering a set-up like this, it is quite unpredictable whether the light will flash or the flag will run up the flagpole!

As we pointed out in Chapter 1, the conclusion of quantum physics that some events are essentially unpredictable contradicts the classical view of physics, which maintains that the behaviour of the universe is governed by mechanistic laws. Classically, it was thought that particles moved under the influence of definite forces and that if all these forces were known, along with the positions and speeds of all the particles at some instant, the subsequent behaviour of any physical system could be predicted. For example, when we toss a coin we think of the outcome as being random, but we know that it is actually determined by the forces acting on it, beginning with the impulse it receives from our hand and ending with its interaction with the floor. Of course such calculations are practicable only in simple cases, but in principle it would be possible to predict the behaviour of any physical system from the laws of physics and the initial conditions. In contrast, whenever we

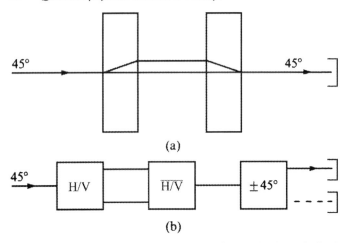

Fig. 2.9 The polarising crystals illustrated schematically in (a) are represented by the boxes H/V and $\overline{H/V}$ in (b). Light split into two components by a calcite crystal can be reunited by a second calcite crystal facing in the opposite direction. If the crystals are set up carefully so that the two path lengths through the apparatus are identical, the light emerging on the right of the $\overline{H/V}$ crystal has the same polarisation as that incident on the left of the H/V crystal. This is also true for individual photons, a fact that is difficult to reconcile with the idea that the photon's state of polarisation is modified when it passes through a polariser (cf. Figure 2.7).

interpret quantum measurement as producing *fundamentally* un-predictable results for experiments such as the measurement of the 45° polarisation of an *h* photon, we are rejecting the traditional deterministic view. Indeterminism and uncertainty are built into the very foundations of the theory so that, in general, the future behaviour of some physical systems cannot be predicted however accurately their present state is known. Recalling Laplace, quantum physics appears to imply that some aspects of the present state of the universe are neither 'the result of its past' nor 'the cause of its future'.

We close this section by introducing a variant on the photon- polarisation measurement that illustrates the key problem of quantum measurement and to which we shall return frequently in later chapters. As shown in Figure 2.9, a 45° polarised light beam is split into its horizontal and vertical components by an H/V polariser in the same way as before, and the two light beams are directed onto a second calcite crystal that faces in the opposite direction. We first consider the case where the light intensity is so high that we can apply the continuous wave model with confidence. One of the properties of polarised waves in calcite crystals is that, because the wavelength of the beam that goes straight through is a little shorter than that of the other, the two light waves emerging from the first crystal have performed an identical number of oscillations and are therefore 'in step'. The same thing happens in the second crystal and the net result is that when the waves are combined, their polarisations also combine (cf. Figure 2.3) to recreate a 45° polarised wave. We can think of this as a type of constructive interference, similar to that discussed in Chapter 1 in the context of the two-slit experiment. It should be noted that, although such an experiment requires careful setting up to ensure that the above conditions are accurately fulfilled, it is a real experiment that can be and often is performed in a modern laboratory.

Now suppose that the light entering the apparatus is so weak that only one photon passes through at any one time. (For the moment we ignore the possibility of hidden variables – see the next section of this chapter.) From what we already know we might be tempted to reason as follows: earlier experiments have shown us that a 45° photon follows only one of the two possible paths through the H/V apparatus. When it reaches the right-hand crystal there is therefore nothing with which it can be reunited, so there is no way in which its 45° polarisation can be reconstructed and it will emerge at random through either channel of the final 45° polariser. When the experiment is performed, however, *the opposite result is observed*: however weak the light, the emergent beam is found to have 45° polarisation, just as the incident beam has. It appears that the superposition state has been maintained throughout after all and that there was no collapse of the photon state after the first polariser. Yet we know that if, instead of the second crystal, we had inserted two detectors, as in Figure 2.8, we would have detected each photon in one or the other of these; in this experiment the photon state collapses in the way we discussed earlier. In the re-unification experiment (Figure 2.9), however, each photon appears to exist in both channels after it emerges from the first polariser. It appears to be the presence of detectors that causes the photon to emerge in one channel or the other, but not both. However, there is no reason why the detectors should not be a long way from each other and from the original polariser – at the other side of the laboratory or even, in principle, the other side of the world. Moreover, the light entering the H/V polariser presumably can have no knowledge of what kind of apparatus (detectors or re-unifying crystal) it is going to encounter on the other side and the possibility that it adjusts its behaviour in the first polariser so as to suit the subsequent measurement is eliminated.

A further possibility may have occurred to some readers. Could it be that the effect of the $\overline{H/V}$ crystal is different from what we thought and that really any horizontally or vertically polarised photon is turned into a 45° photon by such an apparatus? This idea could be tested by blocking off one or other of the beams emerging from the H/V crystal, so that the light entering the crystal $\overline{H/V}$ is certainly either horizontally or vertically polarised. But, whenever this is done we find that the initial polarisation has indeed been destroyed and the photons emerge at random through the two channels of the final 45° polariser. We seem to be forced to the conclusion that the photon either passes along both H/V channels at once (despite the fact that we can only ever detect it in one channel) or, if it passes along only one path, it somehow 'knows' what it would have done if it had followed the other! This is another example of the strange consequences of quantum theory that we encountered in our discussion of two-slit interference in the last chapter.

Hidden variables

Our discussion so far has all been in the context of the conventional interpretation of quantum physics, but there are alternative ways of looking at quantum behaviour. One is to postulate that quantum particles possess hidden properties in addition to those that can be and are observed. Thus, although we cannot know in advance what the outcome of a ±45° measurement on an h photon will be, this does not necessarily mean that it is undetermined. Despite what we said earlier, information on what is going to be the outcome of a future measurement could be a property of the photon, in the same way that the outcome of a coin toss is a property of the state of the coin and the forces acting on it. However, in the quantum case, although the photon may 'know' what it is going to do, such information is always unavailable to (i.e. hidden from) an observer. Theories that follow this approach are therefore known as 'hidden variable' theories.

We now have two different ways of looking at quantum phenomena. One involves assuming that in regard to individual photons the physical world is essentially indeterministic – the photons don't know what they're going to do – while the other postulates that they do know, but conspire to keep this information hidden from us. It may be that each approach makes identical predictions for any known experiment, so that the difference between the theories would be philosophical or metaphysical and undetectable. In this section, we discuss the hidden-variable approach in a little more detail, and we close the chapter with an indication of some of the problems to which hidden-variable theories give rise, some of which will be addressed in subsequent chapters.

Consider again the experiments illustrated in Figures 2.8 and 2.9, taking a hidden-variable point of view. We attribute to the +45° photon some property that decides in advance through which H/V channel it is going to pass while at the same time retaining a memory of its history as +45° photon.

We are now free to assume that the photon after all follows one of the two paths through the polariser and that, if there are detectors in the paths as in Figure 2.8, one of these is activated. If, however, we bring the two paths together and measure the ±45° polarisation, as in Figure 2.9, the hidden variable ensures that it is always detected in the positive channel.

A slightly different form of hidden-variable theory is based on another interpretation of wave–particle duality. Instead of treating the wave and particle models as alternatives, this theory proposes that both are present simultaneously in a quantum situation. The wave is no longer directly detectable, as the electromagnetic wave was thought to be, but has the function of guiding the photon along and adjusting its polarisation. For this reason it is often described as a 'pilot' wave. Thus the 45° polarised photon approaches the H/V polariser and sees the pilot wave being split into two parts. It is directed by the wave into one or other path and has its polarisation adjusted to fit the wave in that path. The same thing happens when it passes through the reunifying crystal: the outgoing wave has 45° polarisation (because it is an addition of the H and V waves in step) and this property is transmitted to the outgoing photon. Again, hidden variables have been used to provide a realistic description of the observed results.

Hidden-variable theories similar to the second described above were first suggested by de Broglie in the 1930s and were developed by a number of workers, notably the physicist David Bohm in the 1950s and 60s. They have been developed to the stage where most of the results of conventional quantum physics can be reproduced by a theory of this kind. However, hidden-variable theories have their own problems, which many people think are just as conceptually unacceptable as those of the conventional quantum approach. These include the fact that the mathematical details of hidden-variable theories are much more complex than those of quantum physics, which are basically quite simple and elegant. In addition, the pilot wave seems to be quite unlike any other wave field known to physics: it possesses no energy of its own yet it is able to influence the behaviour of its associated particles, which in turn have no effect on it. A further problem for hidden-variable theories is that they do not generally preserve *locality*. By this we mean that the behaviour of the photon should be determined by the hidden variables carried along with it or in the wave at the point where the photon happens to be. The hidden-variable model applied to measurements of the type discussed in this chapter does appear to preserve locality, but it turns out that it is unable to do so in all circumstances. In particular, some situations involving the quantum behaviour of pairs of photons turn out to be inexplicable using any local hidden-variable theory. When this problem was first identified, it was realised that no experimental tests of this correlated behaviour of photon pairs had been carried out, so it was possible that quantum physics is actually wrong in such situations and some form of local hidden-variable theory is correct. This possibility has been the subject of considerable theoretical and experimental investigation. Because of its importance and the light it sheds on our general understanding of quantum phenomena, the next chapter is devoted to a reasonably detailed explanation of this work.

3 · What can be hidden in a pair of photons?

Albert Einstein's comment that 'God does not play dice' sums up the way many people react when they first encounter the ideas discussed in the previous chapters. How can it be that some future events are not completely determined by the way things are at present? How can a cause have two or more possible effects? If the choice of future events is not determined by natural laws does it mean that some supernatural force (God?) is involved wherever a quantum event occurs? Questions of this kind trouble many students of physics, though most get used to the conceptual problems, say 'Nature is just like that' and apply the ideas of quantum physics to their study or research without worrying about their fundamental truth or falsity. Some, however, never get used to the, at least apparent, contradictions. Others believe that the fundamental processes underlying the basic physics of the universe must be describable in deterministic, or at least realistic, terms and are therefore attracted by hidden-variable theories. Einstein was one of those. He stood out obstinately against the growing consensus of opinion in the 1920s and 30s that was prepared to accept indeterminism and the lack of objective realism as a price to be paid for a theory that was proving so successful in a wide variety of practical situations. Ironically, however, Einstein's greatest contribution to the field was not some subtle explanation of the underlying structure of quantum physics but the exposure of yet another surprising consequence of quantum theory. This arises from analysis of the quantum behaviour of systems containing two or more particles that interact and move apart. Einstein was able to show that, in some circumstances, quantum physics implies that there is a non-local interaction between separated particles: i.e. they appear to influence each other when far apart, even when there is no known interaction between them (see the discussion at the end of Chapter 2). To avoid this, Einstein concluded that each particle must carry with it an additional 'element of reality' (what we might now call a hidden variable). Ironically, later work has shown that any hidden-variable theory that reproduces the results of quantum physics actually has to be non-local. The arguments underlying these conclusions are discussed in this chapter.

Einstein's ideas were put forward in a paper written with his co-workers, Boris Podolski and Nathan Rosen, in 1935 and, because of this, the topic is often denoted by their initials 'EPR'. Their arguments were presented in the context of wave–particle duality, but in 1951 David Bohm showed that

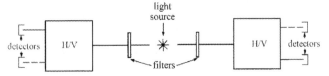

Fig. 3.1 In some circumstances atoms can be made to emit a pair of photons in rapid succession. The two members of each pair move away from the source in different directions; as they have different wavelengths, they can be identified by passing the light through appropriate filters. In the set-up shown, the right-hand apparatus records the polarisation of one of the two photons while that of the other is registered on the left. Whenever a right-hand photon is recorded as horizontally polarised that on the left is found to be vertical, and vice versa.

the point could be made much more clearly by considering measurements of variables, like photon polarisation,[1] whose results were limited to a small number of possible values.

To understand the EPR problem, we focus on a physical system consisting of atoms in which a transition occurs from an excited state to the ground state (see Chapter 1) with the emission of two photons in rapid succession. The wavelengths of the two photons are different, so they correspond to two different colours, say red and green, but their most important property is that their polarisations are always at right angles. If the red photon is vertically plane polarised then the green photon is horizontally plane polarised or if one is polarised at $+45°$ to the horizontal the other is at $-45°$ and so on. Of course not all atoms that emit photons in pairs do so in this way, but some do and experiments on such systems are perfectly practicable, as we shall see. Why do we believe that the polarisations are always at right angles? One answer might be that the quantum theory of the atom requires it, but a more important reason is that this property can be directly measured. Consider the arrangement shown in Figure 3.1. A source consisting of atoms that emit pairs of photons is placed between two filters, one of which transmits red light and the other green light. Each of these light beams is directed at an H/V polariser of the type described in the last chapter, and the outputs from the two channels of each polariser are monitored by photon detectors. The intensity of the emitted light is arranged to be low enough, and the detectors

---

[1] Bohm's paper actually related to the measurement of the angular momentum or 'spin' of atoms. However, it turns out that both the experimental measurements of and the theoretical predictions about particle spin are practically identical to those relating to photon polarisation.

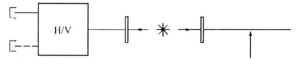

Fig. 3.2 Since we know the polarisations of the pair of photons are always at right angles, the right-hand apparatus of Fig. 3.1 is unnecessary. Whenever a left-hand photon is detected as vertical, we can conclude that its partner (passing the point marked by the arrow) must be horizontal and vice versa. But if a quantum measurement alters the state of the object measured how can the left-hand apparatus affect the polarisation of the right-hand photon many metres away?

operate fast enough, for the pairs of photons to be individually detected. When the apparatus is switched on, it is found that every time an *h* photon is detected in the left-hand polariser, a *v* photon is found on the right and vice versa. There is, of course, nothing special about the H/V configuration, so if both polarisers are rotated through the same arbitrary angle, the same result is obtained. For example if they are set up to make ±45° measurements, a left-hand +45° photon is always accompanied by a right-hand −45° photon and vice versa. Photon pairs created in a state of this kind are said to be 'entangled'.

This may seem quite straightforward, but now consider the set-up in Figure 3.2. This is the same as that just considered except that we have removed the right-hand polariser and detectors because they are unnecessary! If the right-hand polarisation is always perpendicular to that on the left then a measurement of the polarisation of (say) the left-hand photon immediately tells us the right-hand photon's polarisation. That is, the act of measuring the polarisation on the left also constitutes a measurement of the polarisation on the right. But just a minute! This would be perfectly all right if we were making a conventional classical measurement, but we saw in the previous chapter that an important feature of any quantum measurement is that it often affects the state of the system being measured. We don't know what the polarisation of the left-hand photon actually was before it was measured, but it is very improbable that it was exactly horizontal or vertical. The polariser has presumably turned round the polarisation direction of the left-hand photon so that it is *h* or *v* at random; how can it have affected the right-hand photon, which is on the other side of the laboratory by the time the left-hand measurement is made?

If we reject the idea of the measuring apparatus affecting a photon at a distance, how might we explain the fact that the photons are always detected with perpendicular polarisations? One possibility would be to abandon the conventional quantum idea that measurement is a random, indeterministic process and to consider a deterministic hidden-variable theory (see Chapter 2). This would imply that the outcome of any measurement would be determined by the hidden properties of the photons. Each photon would then carry information prescribing what it is going to do when it interacts with a polariser in any orientation. Included in this would be the requirement that, whatever their actual polarisation before they were detected, the members of a photon pair of the type discussed above would always emerge in opposite channels of the two perpendicularly oriented polarisers. It seems then that the properties of photon pairs provide strong evidence for a deterministic hidden-variable theory, of the type discussed in Chapter 2. The outcome of the photon-polarisation measurement would be determined in advance and each photon would 'know what it is going to do' before it enters a polariser.

Such a conclusion is similar to that reached by Einstein and his co-workers in their original paper, whose title is 'Can a quantum-mechanical description of physical reality be considered complete?' If the left-hand apparatus cannot affect the state of the right-hand photon then the set-up in Figure 3.2 must measure some property of the right-hand photon without disturbing it. Even if this property is not the polarisation itself, it must be related to some hidden variable that would determine the result of a polarisation measurement. This quantity must therefore be 'real'. As Einstein puts it:

If, without in any way disturbing the system, we can predict with certainty (i.e. with probability equal to unity) the value of a physical quantity, then there exists an element of physical reality corresponding to this physical quantity.

For the above argument to be generally valid, however, we must suppose that it is possible to construct a local hidden-variable theory that would explain all the experiments we could perform on photon pairs. We would have to consider not only the case where the two polarisers are at right angles but more general situations – such as where one polariser is H/V while the other is ±45°. When we do this, we shall find that no *local* hidden-variable theory is capable of reproducing the predictions of quantum physics in all such cases. It is then up to experiment to decide whether the quantum prediction or those of local hidden-variable theories correspond to reality.

The general situation that we shall consider is one in which we measure the H/V polarisation of the left-hand photon, as before, but now measure the polarisation direction of the second photon parallel and perpendicular to a direction at some angle ϕ to the horizontal, as in Figure 3.3. We will first analyse this experiment using conventional quantum theory and suppressing any doubts we may have about action at a distance. From this point of view, once the polarisation of the left-hand photon has been measured by the H/V

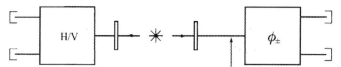

Fig. 3.3 The predictions of quantum physics for photon pairs can be tested by measuring the H/V polarisation of one photon and the $\phi_\pm$ polarisation of the other (i.e. whether it is polarised parallel or perpendicular to a direction making an angle ϕ to the horizontal). If the left-hand measurement puts the right-hand photon into a particular H/V state, the probabilities of subsequent $\phi_\pm$ measurements can be calculated.

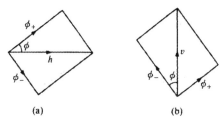

(a)   (b)

Fig. 3.4 If the left-hand photon is vertically polarised we conclude from quantum theory that the one on the right is horizontally polarised before the $\phi_\pm$ measurement. If the $\phi_\pm$ polarisation of $N/2$ such photons is measured, the number recorded as ϕ_+ will be proportional to the square of the ϕ_+ component of the electric field of a horizontally polarised wave. It follows from (a) that this number, $n(v, \phi_+)$, must equal $(N/2) \cos^2\phi$ while the number emerging in the ϕ_- channel, $n(v, \phi_-)$, equals $(N/2) \sin^2\phi$. Similarly it follows from (b) that $n(h, \phi_+) = (N/2) \sin^2\phi$ and $n(h, \phi-) = (N/2) \cos^2\phi$.

polariser to be h or v, the right-hand photon is put into a state of being v or h respectively. We imagine this experiment repeated a large number (say N) times, after which we would expect about $N/2$ photons to have passed through each channel of the H/V polariser. It follows that corresponding to every v photon detected on the left-hand side there is a photon on the right that is horizontally polarised before it reaches the right-hand apparatus. We can calculate the number of these that would be expected to emerge in the positive channel (i.e. polarised at an angle ϕ to the horizontal) of the ϕ apparatus. We call this number $n(v, \phi_+)$ and, from Figure 3.4 and the general theory discussed in Chapter 2, we see that this is just equal to $(N/2) \cos^2\phi$. Similarly, the number emerging in the other channel, $n(v, \phi_-)$, is just $(N/2) \sin^2\phi$. We can apply similar arguments to photon pairs whose left-hand members are horizontally polarised, and we summarise our results as follows:

$$n(v, \phi_+) = (N/2) \cos^2 \phi,$$
$$n(v, \phi_-) = (N/2) \sin^2 \phi,$$
$$n(h, \phi_+) = (N/2) \sin^2 \phi,$$
$$n(h, \phi_-) = (N/2) \cos^2 \phi.$$

We notice in passing that the total number in the positive ϕ channel, $n(v, \phi_+) + n(h, \phi_+)$, is the same as the total number in the negative ϕ channel, $n(v, \phi_-) + n(h, \phi_-)$, and equals $N/2$. This is what we would expect, because we have assumed that the absolute polarisation direction of the photon pair emitted by the source is randomly oriented.

Bell's theorem

The challenge for the hidden-variable approach is to develop a theory of this type that will reproduce the quantum predictions set out above. Referring to the principles of hidden-variable theories set out towards the end of Chapter 2, we see that this means that each photon is ascribed a set of properties that will determine what it will do in any experimental situation. For example, a photon that would be measured as h by an H/V polariser and $+45°$ by a $\pm45°$ polariser has all this information hidden within it before and after either measurement is made. We also wish to assume locality, which means that none of the labels ascribed to one member of a photon pair can be affected by any actions performed only on the other. It will turn out that this is impossible. In 1964 John S. Bell, a physicist working on the properties of elementary particles in CERN near Geneva, showed that no hidden-variable theory that preserves locality and determinism is capable of reproducing the predictions of quantum physics for the two-photon experiment. This was a vitally important theoretical deduction that has been the mainspring of most of the theoretical and experimental research in this field for the last 40 years. In consequence, we devote this section to a proof of what has become known as Bell's theorem. A reader who is uninterested in mathematical proofs and who is prepared to take such results on trust could proceed directly to the conclusions of this section on p. 46 and still keep track of the main argument in this chapter.

We begin by leaving the world of quantum physics to play the following game. Take a piece of paper and write down three symbols, each of which must be either '+' or '−', in any order; write another three plusses and minuses under the first three and repeat the process until you have about 10 or 20 rows each containing three symbols. An example of the kind of pattern generated is given in the following table.

h	ϕ	θ
+	+	−
+	−	+
−	+	−
+	−	+
−	+	+
−	−	+
+	−	−
−	+	−
+	+	+
−	+	−
−	−	+
+	+	−
+	−	+

Label the three columns h, ϕ and θ as shown. Now go through your list and count how many rows have '+' in both column h and column ϕ and call this number $n(h = +, \phi = +)$; in the example above it equals 3. Next count how many have '−' in column ϕ at the same time as '+' in column θ and call this $n(\phi = -, \theta = +)$; it equals 5 in the example. Finally obtain $n(h = +, \theta = +)$, the number of pairs with '+' in both the h and θ columns (4 in the above example). Add together the first two numbers and then subtract the third from your total. If you have followed the instructions properly you will find that your answer is always greater than or equal to zero. That is,

$$n(h = +, \phi = +) + n(\phi = -, \theta = +) - n(h = +, \theta = +) \geq 0.$$

In the particular example this reduces to $3 + 5 - 4 \geq 0$, which is obviously true. Try again if you like and see if you can find a set of triplets composed of plusses and minuses that does not obey this relation; you will not succeed, because it is impossible.

To prove that the above relation must hold for all sets of numbers of the type described, refer to the pie chart in Figure 3.5. As each of the quantities h, ϕ and θ can adopt one of two values (+ or −) all the rows in the table must be contained in one of the eight slices of pie, while the size of a slice represents the number of times the corresponding line appears in the table. It should now be clear that all the quantities in the inequality above are

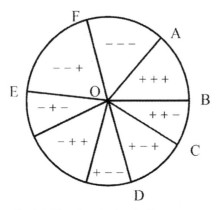

Fig. 3.5 The slices in the pie diagram represent the numbers of triplets of the types indicated. The plusses and minuses refer to h, ϕ and θ in that order.

represented by combining two slices of the pie. Thus

$$n(h = +, \phi = +) = \text{slice AOB} + \text{slice BOC},$$
$$n(\phi = -, \theta = +) = \text{slice COD} + \text{slice EOF},$$
$$n(h = +, \theta = +) = \text{slice AOB} + \text{slice COD}.$$

Referring to Figure 3.5, we see that the left-hand side of the inequality is just the sum of the slices BOC and EOF, which must represent a positive number (or zero), and this is the required result.

What has all this got to do with the properties of pairs of polarised photons? Imagine first that we could measure the polarisation component of a photon in three different directions, without disturbing it in any way. We could then find out whether it was: (i) parallel or perpendicular to the horizontal; (ii) parallel or perpendicular to a direction at ϕ to the horizontal and (iii) parallel or perpendicular to a third direction at θ to the horizontal. If in each case we wrote down a plus when the result was parallel to the given direction and a minus when it was perpendicular, and if we repeated the experiment a number of times, we would get a set of numbers just like those in the table, which would therefore be subject to the above relation.

Of course we can't do this directly because it follows from the discussion in Chapter 2 that it is not possible to make simultaneous measurements of two or more polarisation components of a single photon. However, if we assume that a local hidden-variable theory holds, we can obtain values for any two out of the three components by making measurements on entangled photon pairs, remembering that the polarisations of the two members of any pair are always perpendicular. For example, referring back to Figure 3.3,

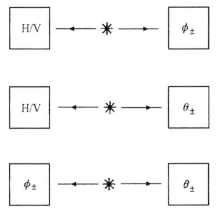

Fig. 3.6 In proving Bell's theorem we consider three separate sets of measurements on entangled photon pairs, in which the polarisers are set in the orientations indicated.

if one member of a pair is measured as v (i.e. $h = -$) on the left while the other is found to be $\phi = +$ on the right, then we can conclude that the right-hand photon belonged to the set $(h = +, \phi = +)$. Now consider three separate experiments as set out in Figure 3.6; each experiment is performed on a different set, each set containing the same number of similarly prepared photon pairs. In the first, H/V (i.e. $h = +$ or $h = -$) is measured on the left and $\phi_\pm$ is measured on the right; in the second, H/V is measured on the left and $\theta_\pm$ on the right, while in the third $\phi_\pm$ and $\theta_\pm$ are measured. Considering the first experiment, if the polarisation of the left-hand photon is found to be v then if we had measured H/V on the right we would certainly have found that the right-hand polarisation was $h = +$. But on the right we have actually measured the polarisation in the $\pm\phi$ direction. It follows that the number of pairs – $n(v, \phi_+)$ – where the polarisation of the left-hand photon is vertical and that of the right-hand photon is parallel to the ϕ direction is equal to $n(h = +, \phi = +)$ as defined above and applied to the right-hand photons. That is,

$$n(v, \phi_+) = n(h = +, \phi = +).$$

The second and third sets of experiments give us in exactly the same way

$$n(v, \theta_+) = n(h = +, \theta = +)$$

and

$$n(\phi_+, \theta_+) = n(\phi = -, \theta = +).$$

It follows directly from these relations and the conclusion we drew from the game we played earlier that we can make a prediction about the results of such a set of measurements:

$$n(v, \phi_+) + n(\phi_+, \theta_+) - n(v, \theta_+) \geq 0$$

This is a version of what is known as Bell's inequality. Putting it into words, it states that:

1. *If* we carry out three experiments to measure the polarisations of three sets of equal numbers of photon pairs, where the polarisation directions of the right-and left-hand polarisers are: in the first experiment, horizontal and at an angle ϕ to the horizontal; in the second vertical and at an angle θ to the horizontal; and in the third at angles ϕ and θ to the horizontal respectively:

2. *Then* the total number of pairs in which the left- and right-hand photons are detected as vertical and positive respectively in the first experiment and as doubly positive in the third can never be less than number of doubly positive pairs in the second experiment;

3. *Provided that* the results of the experiments are determined by hidden variables possessed by the photons and that what happens to one photon is unaffected by the other or by the setting of the distant apparatus.

An important point to note is that the three experiments are separate – i.e. they are performed on different sets of photon pairs. An underlying assumption, implied in our assumption of a hidden-variable theory, is that if we had measured, say, $n(v, \theta_+)$ for set one (for which we actually measured $n(v, \phi_+)$) we would have obtained the same result as we did in the actual measurements performed on set two (assuming that the numbers of pairs in each set are large enough for statistical differences to be small). This is an example of what is known as 'counterfactual definiteness', by which we mean that it is meaningful to talk about events that did not take place, but *would have occurred* if circumstances had been different. Counterfactual definiteness is a clear consequence of a local deterministic hidden-variable theory, but it can be a moot point in the context of other interpretations of quantum physics.

Let us now see whether quantum theory is consistent with Bell's inequality. The quantum theory expression for $n(v, \phi_+)$ is given above as $(N/2) \cos^2 \phi$ and $n(v, \theta_+)$ is similarly $(N/2) \cos^2 \theta$. To obtain an expression for $n(\phi_+, \theta_+)$ we see that if we rotate our axes through an angle ϕ then the direction ϕ_+ corresponds to the new 'horizontal' direction and the direction θ_+ is at an angle $(\theta - \phi)$ to this. Hence $n(\phi_+, \theta_+) = n(h, (\theta - \phi)_+) = (N/2) \sin^2(\theta - \phi)$. It follows that Bell's inequality and quantum physics could be consistent if

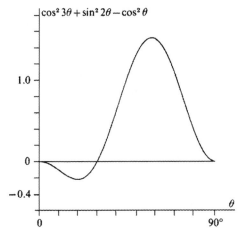

Fig. 3.7 If quantum physics were consistent with Bell's theorem, then the function $\cos^2 3\theta + \sin^2 2\theta - \cos^2\theta$ would have to be positive or zero for all values of θ. The graph shows that this is not true for values of θ between $0°$ and $30°$.

and only if

$$\cos^2\varphi + \sin^2(\theta - \varphi) - \cos^2\theta \geq 0$$

for all possible values of θ and ϕ. On the other hand, to show that quantum physics is inconsistent with Bell's inequality it is only necessary to show that this relation is false for some particular values of θ and ϕ. It is convenient to consider the special case where $\phi = 3\theta$, when the left-hand side becomes $\cos^2 3\theta + \sin^2 2\theta - \cos^2\theta$. Figure 3.7 shows this expression as a function of θ, and we see that, although Bell's theorem is obeyed for values of θ greater than $30°$ it is clearly violated between $0°$ and $30°$. In the particular case for which $\theta = 20°$ and $\phi = 60°$, the expression equals -0.22 in clear breach of Bell's inequality. We are now in a position to state Bell's theorem, which is:

No local hidden-variable theory can reproduce the predictions of quantum mechanics for all experiments.

We are therefore forced to the conclusion that either quantum physics does not correctly predict the results of polarisation measurements on photon pairs or one of the assumptions on which Bell's theorem is based is wrong. But these are extremely basic assumptions. We have simply said that the outcome of a photon-polarisation measurement cannot be affected by the way in which another distant apparatus (typically several metres away) is set up and that the results of such a measurement are determined by some hidden property of the photon. The only other assumptions are some basic rules of logic and mathematics! The first alternative – that quantum physics is wrong in the case of photon pairs – can in principle be tested by experiment, so it is hardly surprising that the appearance of Bell's theorem in the late 1960s stimulated considerable experimental effort in this field.

The experiments

It might be thought that the outstanding success of quantum theory across the whole compass of physical phenomena would have meant that the predictions of Bell's theorem would already have been tested implicitly by experiments performed before its publication in 1969. For example, the excellent agreement between the calculated and observed properties of helium, whose atoms each contain two electrons, might be thought to be sensitive to the properties of entangled pairs of particles. However, it soon became apparent that no direct test of the particular kind of correlation involved in the Bell inequality had actually been made. There have been a number of cases in the development of physics where it has been wrongly assumed that a theoretical statement has been experimentally tested, and when an actual experiment has been done this has shown that the theory accepted until then was wrong. (One example of this was the discovery in the 1950s that some processes in particle physics depend on the parity (i.e. right- or left- handedness) of the system.) Thus, even though few physicists doubted that quantum physics would turn out to be correct, it was very important that a direct test of Bell's theorem be made.

It was soon realised, however, that there are severe practical difficulties preventing a direct experimental test of Bell's inequality in the form given above. These arise particularly because neither polarisers nor photon detectors are ever 100 per cent efficient, so that many of the photons emitted by the source are not actually recorded. Moreover, these inefficiencies can depend on the setting of the polarisers, so rendering the proof in the previous section inapplicable to the practical case. Further consideration of these problems led to the derivation of a new form of Bell's theorem that is not subject to this criticism, though it does depend on some additional assumptions. These include the apparently obvious assumption that passing light through a polariser can only diminish its intensity and never enhance it. The Bell inequality generalised in this way involves the consideration of an experiment in which measurements are made with four relative orientations of the polarisers instead of the three settings considered above. We shall not discuss the proof of this, but simply state the result:

$$n(v, \phi_+) - n(v, \psi_+) + n(\psi_+, \phi_+) + n(\theta_+, \psi_+) \leq n(\theta_+) + n(\phi_+)$$

where the left-hand polariser can be set to measure either H/V polarisation or polarisation at an angle θ to the horizontal and the two orientations of the right-hand polariser are φ and ψ. The numbers on the left-hand side of the above inequality refer to the number of times photons are simultaneously recorded in the appropriate channels, while the numbers on the right-hand side refer to measurements made with one of the polarisers absent: thus $n(\theta_+)$ represents the number of times a photon appears in the left-hand θ_+ channel at the same time as a right-hand photon is recorded with no right-hand polariser present, and $n(\phi_+)$ is defined similarly for the case when the left-hand polariser is removed. It can be seen that six separate experiments are required to collect all the relevant data, and it is assumed that the same total number of photon pairs is involved in each case.

The extended Bell inequality has a further important property, which we shall also state without proof. This is that it tests not only deterministic theories but also a wide range of hidden-variable theories that include an element of randomness, such as some variants of the pilot-wave model referred to near the end of Chapter 2. If the extended inequality is breached, any kind of hidden-variable theory that preserves locality is ruled out. A number of experiments testing Bell's theorem in this or similar contexts have been carried out since 1970. Although one of the early experiments initially produced results consistent with Bell's theorem and in disagreement with quantum theory, all the others (including a repeat of the experiment just referred to) agree with quantum predictions and breach the Bell inequality. Some of the early experiments were carried by John Clauser and Abner Shimony, among others, and the most definitive of those performed during the fifteen years or so following the discovery of Bell's theorem were carried out by Alain Aspect in France. In particular, his 1982 experiment was of the type described above.

When his experimental results were substituted into the above expression, it was found that the left-hand side, instead of being smaller than the right-hand side, was actually larger by an amount equal to $0.101N$, where N is the total number of photon pairs in each run. A quantum calculation of the same quantity (taking detection efficiency into account) produces the result $0.112N$. The estimated errors in the experiment were large enough to embrace the quantum result but small enough to rule out the Bell inequality.

The Aspect experiment possessed an additional feature of considerable interest. This is illustrated in Figure 3.8, where we see that the two polarisation measurements on each side were actually made using different polarisers, which were always in position, the photons being switched into one polariser or the other by the devices marked S in order to perform the appropriate measurements. These 'switches' are in fact actuated by high-frequency ultrasonic waves, so that the measurement is switched from one channel to another about one hundred million times per second! This rapid

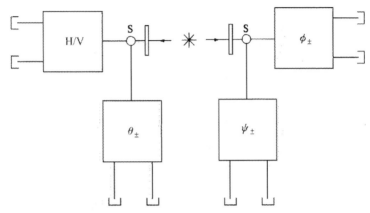

Fig. 3.8 In the Aspect experiment to test the predictions of Bell's theorem for the polarisation of photon pairs, either the H/V or the $\theta_\pm$ polarisation can be measured on the left and either $\phi_\pm$ or $\psi_\pm$ on the right. The ultrasonic switches, S, operate so quickly that it is impossible for the photons to be influenced by the settings of the distant apparatus unless such influences can be propagated faster than light. The results of this experiment agree with the predictions of quantum physics and are inconsistent with any local hidden-variable theory.

switching has an important consequence. Remember that at the beginning of our discussion (see Figure 3.3) we showed how the quantum result follows from the fact that the setting of the left-hand apparatus apparently influences the state of the right-hand photon. We might envisage, therefore, that the left-hand apparatus is sending some kind of 'message' to the right-hand photon telling it how it is set up, so that the right-hand photon can interact in an appropriate way with the right-hand polariser. In the Aspect experiment with switching at the above rate, any such message must travel in a time less than one hundred millionth of a second – otherwise the right-hand photon would behave as if it had been measured by a polariser set in an earlier orientation of the left-hand apparatus. But the two polarisers are about 10 metres apart and a signal travelling at the speed of light would take about three times the switching time to cover this distance. As it is well known that no physical object or signal can travel faster than light, the possibility that the right-hand photon is receiving messages from the other detector must be discounted.

Experimental tests have been extended in recent years to the point where the sets of measuring apparatuses are separated by several kilometres, so that the speed of any signalling would have to be many times the speed of light. Different types of sources of photon pairs have been developed and experiments have also been performed on entangled triplets. In every case the results have agreed with the predictions of quantum physics and have breached Bell's theorem to a greater or lesser extent. This work has had a spin-off in that it has increased general interest in the properties of entangled quantum systems, with possible practical applications to areas such as cryptography and computing.

Considerable thought has gone into a critical appraisal of the design details of experiments such as those described above to see whether the results are really inconsistent with Bell's theorem or whether some loophole in the reasoning remains. The only conceivable possibility seems to be that the efficiencies of the photon detectors might somehow depend on the hidden variables and in some circumstances actually act as amplifiers – so breaching the 'no enhancement' assumption mentioned above. This would not of course be possible if the detectors always registered every photon entering them because extra photons would have to appear from nowhere, but real detector efficiencies are quite low, so that this loophole cannot be said to be completely closed by experiments on photon pairs. However, experiments in which the detection efficiency is essentially 100% were carried out on correlated pairs of atoms early in the twenty-first century. Once again the quantum predictions were confirmed and Bell's theorem falsified, although in these experiments there was no rapid switching between the detectors, so communication between the atoms could not be ruled out. Nevertheless, even if some small loophole remains, the fact that all the experimental results are not only inconsistent with Bell's theorem but also agree with the quantum predictions must motivate us to look for something other than local hidden-variable theories to resolve the puzzles of quantum physics.

Discussion

It should now be clear why the results of two-photon experiments lead some people to describe the Einstein–Podolski–Rosen effect as a 'paradox'. Although some scientists strongly disagree with this label, there is something almost paradoxical about the position in which we find ourselves. We started off by showing that quantum physics implies that an operation carried out on one photon appears to affect the state of another a long distance away. We then considered the alternative possibility that the effect arises from some hidden property possessed by each photon when the pair is created, but Bell's theorem and the associated experiments forced us to reject this. Now the final Aspect experiment proves that no message can be passed to a photon from the distant apparatus unless it travels faster than the speed of light. At first sight this seems to contradict the initial statement that we thought had just been confirmed!

There is certainly no easy answer to the problem of non-locality and the EPR paradox. In the next chapter we shall return to this question in the context of a deeper discussion of the conventional or 'Copenhagen' interpretation of quantum theory, but for the moment we make several points that should be kept in mind in any discussion of this question.

First, although the properties of the photon may appear to depend on the settings of the distant apparatus, this is not the sort of influence we are used to encountering between physical objects. In particular, we cannot use it to transmit information or signals from one place to another. To see why this is so, consider two experimenters at opposite ends of a two-photon apparatus trying to use the equipment to pass signals to each other. Assume for the moment that the left-hand apparatus is oriented as H/V and is nearer the light source than is the $\phi_\pm$ apparatus on the right (see Figure 3.3). The experimenter on the left rotates her apparatus, believing that this will affect the possible polarisations of the photons she measures, and also those of their partners on the right. If the experimenter measuring the polarisation of these right-hand photons could deduce from his results what the left- hand settings must have been then a line of communication would have been established. However, this is impossible because, even if the left- hand polariser puts one right-hand photon into a state of (say) vertical polarisation, it will put the next one into an h or v state *at random*. When the $\phi_\pm$ polarisation is measured on this set, we expect an equal number of photons (see the discussion on p. 39) to emerge randomly from the two channels of the right-hand polariser and observing this provides no information about the set-up on the left. An experimenter cannot draw any conclusions about the orientation of the other apparatus by observing only his 'own' photons; it is only when the results of the measurements on both sides are brought together, so that the correlations between them can be examined, that the effects discussed in this chapter are observed.

Second, the fact that it is impossible to use the EPR set-up to transmit information at least partly resolves the problem of influences apparently travelling faster than light. If such influences cannot be used to transmit information then they need not be subject to the theory of relativity, which requires that no *signal* can be transmitted faster than light. We are dealing with a *correlation* between two sets of events, which does not travel in one direction or the other. When we first analysed the situation in which the left-hand apparatus measured H/V polarisation while that on the right made measurements at an angle ϕ to the horizontal (see Figure 3.3), we said that the left-hand apparatus had put the right-hand photon into an h or v state, which was then further analysed by the right-hand polariser. But if we had put it the other way round, so that the right-hand measurement put the left-hand photon into an h or a v state, we would have obtained the same answers for all the experimental results such as $n(v, \phi_+)$ etc. The same would be true if the measurements had been made simultaneously. Readers familiar with the theory of relativity will know that the time order of 'space-like separated' events[2] can be reversed by observing them from an appropriate moving frame of reference. It follows that, provided the measurements are made in rapid succession, an observer travelling past the apparatus in Figure 3.3 from left to right at a sufficiently high speed would conclude that the right-hand photon had been detected before that on the left and therefore that the ϕ measurement had occurred first. However, as the measured correlations do not depend on any assumed direction of travel, all experimental results are unaffected. As a result, it has been said that there is 'peaceful co-existence' between the EPR experiments and the theory of relativity.

The aim of this chapter has been to demonstrate that it is impossible to avoid the revolutionary conceptual ideas of quantum physics by postulating a hidden-variable theory that preserves locality Bell's theorem and experiment have shown that the observed properties of pairs of photons cannot be explained without postulating some correlation between the state of the measuring apparatus and that of a distant photon. It should also be noted that, even if quantum theory were shown to be incorrect tomorrow, any new fundamental theory would also have to face the challenge of the violation of Bell's inequality and would have to predict the observed correlations between widely separated measurements. Nevertheless, we should note that Bell's theorem applies to the local hidden-variable interpretations of quantum physics and that its implications for other approaches are not clear cut. The argument depends on the assumption that a photon 'knows' what it is going to do, but if we reject this and the realistic approach that underlies it, it is not at all clear that non-locality is still inevitable. What is now the orthodox approach to quantum physics adopts this radical point of view, questioning whether the postulate of some correlation between the state of a measuring apparatus and that of a distant photon is meaningful and indeed whether photons can be said to have any existence at all until they are in some sense observed. This viewpoint is known as the Copenhagen interpretation and is the subject of the next chapter.

---

[2] Space-like separated events occur at two different places in such rapid succession that no light signal could pass between them during the time interval separating them.

4 · Wonderful Copenhagen?

The 1935 paper by Einstein, Podolski and Rosen represented the culmination of a long debate that had begun soon after quantum theory was developed in the 1920s. One of the main protagonists in this discussion was Niels Bohr, a Danish physicist who worked in Copenhagen until, like so many other European scientists of his time, he became a refugee in the face of the German invasion during the Second World War. As we shall see, Bohr's views differed strongly from those of Einstein and his co-workers on a number of fundamental issues, but it was his approach to the fundamental problems of quantum physics that eventually gained general, though not universal, acceptance. Because much of Bohr's work was done in that city, his ideas and those developed from them have become known as the 'Copenhagen interpretation'. In this chapter we shall discuss the main ideas of this approach. We shall try to appreciate its strengths as well as attempting to understand why some believe that there are important questions left unanswered.

When Einstein said that 'God does not play dice', Bohr is said to have replied 'Don't tell God what to do!' The historical accuracy of this exchange may be in doubt, but it encapsulates the differences in approach of the two men. Whereas Einstein approached quantum physics with doubts, and sought to reveal its incompleteness by demonstrating its lack of consistency with our everyday ways of thinking about the physical universe, Bohr's approach was to accept the quantum ideas and to explore their consequences for our everyday ways of thinking.

Central to the Copenhagen interpretation is a distinction between the microscopic quantum world and the everyday *macroscopic* apparatus we use to make measurements. The only information we can have about the quantum world is obtained from these measurements, which generally have an effect on the system being measured. It is pointless to ascribe properties to an isolated quantum system, as we can never know what these are until we measure them. According to Bohr, they have no reality in the absence of measurement, as real physical properties are possessed only by the combined system of microscopic object *plus* measuring apparatus.

We can demonstrate these ideas more clearly by again considering the measurement of photon polarisation. Suppose, as in Chapter 2, a photon with unknown polarisation approaches an H/V apparatus and emerges in one of the channels – say the V channel. According to Bohr it is incorrect to

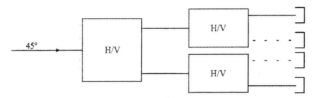

Fig. 4.1 If photons known to be polarised at 45° to the horizontal are passed through an H/V polariser they emerge at random through the horizontal and vertical channels. After this the photons are apparently either horizontally or vertically polarised, as is confirmed by further H/V measurements.

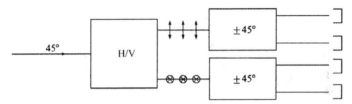

Fig. 4.2 If 45° polarised photons are incident on an H/V polariser they emerge as either horizontally or vertically polarised. If the two beams are then passed through separate ± 45° polarisers, they appear in one or other of the output channels at random.

speculate on what the photon polarisation was before the measurement as this is unknowable. After the measurement, however, it is meaningful to say that the photon is vertically polarised because it has been detected in the V channel, and if we pass it through a second H/V apparatus we know that it will certainly emerge through the V channel again (Figure 4.1). If, however, we direct the vertically polarised photon towards a polariser set at some other angle, say at 45° to the horizontal, then we cannot know in advance through which ±45° channel the photon will emerge (Figure 4.2). The Copenhagen interpretation then states that it is incorrect to attribute any reality to the idea that the photon possesses 45° polarisation before this is measured. Moreover, after the 45° measurement has been made, any knowledge of the previous h or v state will have been destroyed and it is then incorrect to attribute *this* property to the photon.

Bohr described the fact that a measurement of one property of a quantum system generally destroys all knowledge of some other property as 'complementarity'. Thus H/V and ±45° polarisations are referred to as complementary variables, whose values could never be simultaneously measured and therefore can never be simultaneously ascribed to a photon. Certainly, to describe a classical wave as being simultaneously vertically polarised and polarised at 45° to the horizontal would be a contradiction in terms, and it would be natural to apply the same logic to the individual photons. If we do so, we are almost inevitably led to the concept of indeterminism, because if it is meaningless to ascribe a particular 45° polarisation to a vertically polarised photon then there is nothing to determine the result of a subsequent 45° measurement. Bohr willingly embraces the fundamental indeterminism of quantum physics and, rather than trying to recover a mechanistic model through some form of hidden-variable theory, he treats complementarity and indeterminism as fundamental facts of nature, which our studies of quantum physics have led us to appreciate.

The wave analogy may make the complementary nature of different photon polarisation directions appear reasonably acceptable, but the application of the idea to other physical systems requires a much more radical change in our thinking. Thus in Chapter 1 we showed how it is impossible to make simultaneous precise measurements of the position and momentum of an electron: measurement of one quantity inevitably renders unpredictable the result of a subsequent measurement of the other. In this case the Copenhagen interpretation says that it is not meaningful to think of the electron as 'really' possessing a particular position or momentum unless these have been measured. Moreover, if its momentum (say) has been measured it is then meaningless to say that it is in any particular place. So also with wave–particle duality (see Chapter 1): when light or an electron beam passes through a two-slit apparatus it behaves as a wave because in these circumstances it *is* a wave. However, when it is detected by a photographic plate or a counter, it behaves like a stream of particles because in this context it *is* a stream of particles. The object and the measuring apparatus *together* determine the possible outcomes of a measurement: we must not ascribe properties to the object alone unless and until these have been measured.

We might wonder how we know that a quantum object exists at all in the absence of any measurement. The Copenhagen answer is that sometimes we don't. Until we have measured some property of a system it is meaningless to talk about its existence. When a property has been measured, however, it is meaningful to talk about the existence of the object with this property until some complementary property has been measured. Thus it is usually meaningful to attribute a definite mass and charge to an electron whose existence is thereby established, and most subsequent measurements will leave these properties unaltered. In some circumstances, however, such as when an electron collides with and annihilates a positron to produce two γ-ray photons, even these quantities change in a quantum manner and lose their reality.

Copenhagen and EPR

The differences between Bohr and Einstein in their approach to what was then the still new subject of quantum physics led to a lively debate between the two that was conducted at several scientific conferences and in the scientific literature of the time. On many occasions Einstein would suggest a subtle experiment by which it seemed that the values of a pair of complementary variables could be simultaneously measured, and Bohr would reply with a more careful analysis of the problem, showing the simultaneous measurement to be impossible. In 1935, however, came the EPR paper, whose implications we discussed in the previous chapter. This showed how quantum physics requires that a value for a property, such as the polarisation of a photon, could be obtained by measuring the polarisation of a second photon that had interacted with the first some time previously, but was now a long way from it. If it is inconceivable that this measurement could have interfered with the distant object, it appears to follow that the first photon must have possessed the measured property *before* the measurement was carried out. As the experimenter adjusting the distant apparatus can vary the property measured, EPR concluded that all physical properties, in our example values of photon polarisation in all possible directions, must be 'real' *before* they are measured. This is in direct contradiction to the Copenhagen interpretation (remember that this was before Bell's theorem and the experiments discussed in the last chapter).

According to Bohr's colleague Leon Rosenfeld writing in 1967, the EPR paper had been an 'onslaught that came down upon us like a bolt from the blue'. Bohr immediately abandoned all other work to concentrate on refuting the new challenge. Eventually he succeeded (to his own satisfaction at least) and commented to Rosenfeld 'They [EPR] do it smartly, but what counts is to do it right'.

How then did Bohr 'do it right'? How is the Copenhagen interpretation able to resolve the paradoxes described in the last chapter, problems that bothered the minds of physicists thirty years after Bohr, leading to the Bell theorem and the Aspect experiments? The essence of Bohr's reply is that in this example the quantum system consists of two photons, but these must not be considered as separate entities until after a measurement has been made to separate them. It is therefore wrong to say that they have not been disturbed by the left- hand measurement, as it is this that first causes the separation to occur by acting only on the left-hand photon. He also points out that the indirect method of measurement causes no breach in the rules of complementarity because, having chosen to measure (say) the H/V polarisation on the left, it is this same component whose value we have determined on the right-hand photon. We can get a value for a different right-hand component only by performing a further measurement that will cause a further disturbance to the system. Since the result of this further measurement cannot be predicted in advance it is incorrect of EPR to deduce that it must be 'an element of reality' before the final measurement.

We see here the central idea of the Copenhagen interpretation: a quantity can be considered real only if it has been measured or if it is in a measurement situation where the outcome of the experiment is predictable. An experimenter who rearranges her apparatus can change the measurable properties of a quantum system. As Bohr puts it 'there is essentially *the question of an influence on the very conditions that define the possible types of prediction regarding the future behaviour of the system*' (Bohr's italics).

It is well worth pausing at this point to consider the implications of this statement of Bohr's because they go to the very heart of his approach to quantum measurement theory. There are effectively three different levels of operation in a quantum measurement. The first consists of the way the measuring apparatus is set up (e.g. which components of polarisation are being measured by the Aspect polarisers in Figure 3.10). The second level is the statistical result that is obtained after a large number of measurements have been made (e.g. how many photons emerge in each channel), and the third is the result actually obtained in a particular, individual measurement. As far as this last is concerned (apart from special circumstances such as when a previously known polarisation component is re-measured) this is completely random and unpredictable. At the second level, quantum physics allows us to predict future statistical behaviour if the present state is known: thus if a large number of 45° photons pass through an H/V apparatus we know that half of them will appear in each channel. At the first level, the way the apparatus is set up determines what type of property will be measured and therefore, as Bohr says, what 'possible types of prediction regarding the future behaviour of the system' can be made. In the Aspect experiment it is this first level which is changed one hundred million times per second by the ultrasonic switch. As far as measurements on an individual photon pair are concerned the results of these third-level processes are random and unpredictable whatever the setting of the apparatus. The second-level statistical predictions are affected by the first-level changes in a way that can be predicted by quantum theory and Bohr would certainly not have been surprised that these predictions are confirmed by experiment.

With the benefit of the further insights from the work of Bohm, Bell and Aspect discussed in the last chapter, does the Copenhagen interpretation of EPR still hold up? From one point of view, certainly. The experimental results are entirely consistent with the quantum predictions and the 'possible types of prediction' are indeed influenced by the experimental conditions – even when these are altered one hundred million times a second in the Aspect experiment. Bell's theorem relates to the properties of hidden-variable theories, which play no part in the Copenhagen approach, and it was Einstein who (incorrectly) believed that such 'elements of reality' were needed to resolve the problem of non-locality. Does this mean that there is no non-locality in the Copenhagen interpretation? Well, the arrangement of the apparatus affects only the 'possible types of prediction', but it does so *instantaneously* for the *whole* system including the distant photon, so how can this be consistent with locality? The key point here is that Bohr does not assume any physical change in the quantum system upon alteration of the experimental conditions. What have changed are not the 'possible properties' of the system but the 'possible types of prediction' that we are able to make. The change in the apparatus affects only the model that the observer uses to describe the system, not the system itself. Where EPR assume that non-locality is impossible and deduce that the quantum model must be wrong, Bohr assumes that the quantum model is correct and therefore the instantaneous correlations between the results recorded on the separate measuring apparatuses are expected and do occur. Since there are no hidden variables, Bell's proof that these imply non-local interactions does not apply. Non-locality is avoided by eschewing the possibility of making any realistic description of the system between measurements. Difficulties arise only when we try to extrapolate beyond the actual measured reality and to attribute 'reality' to the photons before they interact with the apparatus. The Copenhagen interpretation prohibits this and considers any such unmeasurable properties to be unreal and meaningless.

We see that the Copenhagen interpretation involves a complete revolution in our thought compared with the classical approach and it is this psychological change that Bohr believed is forced on us by the development of quantum physics. Indeed, as was mentioned earlier, most modern undergraduate courses in physics seem to be aimed at conditioning students to think in this unfamiliar way. The very fact that this is difficult can encourage a spirit of conformism to the Copenhagen viewpoint, because students may attribute any doubts they have about it to inadequacies in their understanding, rather than any flaw in the argument. Einstein himself reacted to Bohr's reply with the comment that Bohr's position was logically possible, but 'so very contrary to my scientific instinct that I cannot forego my search for a more complete conception'. So far no such 'more complete conception' has been found and we may have to make the best of the Copenhagen interpretation. It turns out, however, that this leads us to another major problem whose conceptual and philosophical implications probably exceed anything discussed so far. It is this 'measurement problem' that we shall outline in the remainder of this chapter and discuss in the rest of this book.

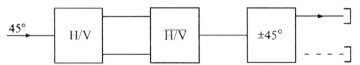

Fig. 4.3 If the photons emerging from the H/V polariser are passed through a reversed polariser the original 45° state can be reconstructed. We must conclude that a quantum measurement of polarisation cannot be performed using an H/V polariser alone.

The measurement problem

To explore the nature of the measurement problem in quantum physics we return again to an example of polarisation measurement that we discussed near the end of Chapter 2. Since it will be very important for our discussion, we outline these arguments again here. Consider a photon, whose polarisation is known from a previous measurement to be at $+45°$ to the horizontal, passing through an H/V polariser such as a calcite crystal. The elementary question we might ask is 'Does the calcite crystal actually measure the photon polarisation?' The obvious reply must be 'yes', and if we ask 'how do we know?' the answer surely is that passage through further polarisers will confirm that the photons emerging in the two channels are horizontally and vertically polarised respectively. In particular, if the photons are passed through further H/V polarisers, all those emerging from the first one in the H or V channel pass through the corresponding channel of the others, as in Figure 4.1.

Now, however, consider the experiment described at the end of Chapter 2 and illustrated again in Figure 4.3. A beam of photons polarised at 45° to the horizontal is passed through a polariser oriented to measure H/V polarisation as before. Now, however, the photons emerging from the two polarisation channels are brought into the same path by a reversed calcite crystal, so that when the final beam is examined it is impossible to tell through which channel a particular photon has passed. If the H/V apparatus has indeed performed the measurement, the emergent beam must be a mixture of horizontally and vertically polarised photons, and if these are now passed through a further $±45°$ polariser we should expect them to emerge at random in the $+45°$ and $-45°$ channels.

In reality, however, this does not happen. Provided that the apparatus is carefully set up, so that the distances travelled along the two paths through the H/V crystals are exactly the same, the photons emerging from the second crystal are all found to be polarised in the $+45°$ direction, just as they were initially. The effect of the supposed 'measurement' has been entirely cancelled out! We do not know through which H/V channel a particular photon passed, so we have not actually measured the H/V polarisation at all. As was explained in Chapter 2, this behaviour is consistent with that expected from the wave model of light. Moreover, even with weak light where the photons pass through the apparatus one at a time, this is a perfectly practicable experiment, which has been performed many times.

Let us consider this situation using the concept of *superposition*. We showed in Chapter 2 that a photon in a 45° polarisation state could be considered as being in a superposition of h and v. It turns out that, if we apply the standard rules of quantum physics to a photon passing between the polarisers in Figure 4.3, its state is similarly predicted to be a superposition of an h state in the H channel and a v state in the V channel. This superposition evolves quite naturally into the 45° state after the photon has passed through the reverse polariser. However, if we had measured its polarisation while it was between the two polarisers, we would not have observed a superposition: instead we would have found that the state had 'collapsed' into *either h or v*. The heart of the measurement problem is the question of what it is that is special about the measurement context, so that we get a random 'either- or' outcome rather than a superposition state.

Thus if an H/V measurement is actually to be performed, the photon must have been *detected* in one channel or the other by a photon counter or similar device. That is, a photon can be considered to be vertically or horizontally polarised only if its passage through the appropriate channel of an H/V polariser has been recorded by such a detector. In the absence of such a record, we should not be surprised that passing the photon through the two polarisers does not change the previous polarisation. We saw earlier how Bohr always stressed the importance of measurement and warned against ascribing reality to unmeasured properties of quantum systems. From the Copenhagen point of view, it is meaningless to attribute the property h or v to the photon emerging from a polariser in the absence of detection. Only when the photon has passed through a polarising apparatus *which includes a detector* should we consider the concept of photon polarisation to be meaningful at all.

Photon detection can mean that the particle has entered some kind of apparatus that has recorded its passage by emitting a click or in some other way. However, detection does not need to be quite so direct. For example, we could block off one of the two paths – say the vertical one – with a shutter. All the emerging photons would then appear in the horizontal channel and would then pass randomly through one or other channel of the subsequent ±45° polariser (Figure 4.4). A similar thing would happen if we blocked off the horizontal channel. Moreover, if we were to set up some mechanical system to move the shutter in and out of each path in turn we would again find that the emerging beam would consist of a mixture of horizontally and vertically polarised photons (cf. an analogous situation in Figure 1.4). We

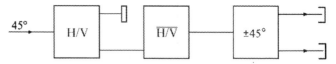

Fig. 4.4 If one of the paths between H/V and $\overline{HV}$ is blocked, the 45°
polarization can no longer be reconstructed and the photons emerge at
random through the two channels of the 45° polariser.

could try something subtler and, instead of using a shutter, put some kind of
photon detector in the beam that still allows the photon through. We might
then expect to be able to record the H/V polarisation without destroying the
final ±45° polarisation. However, it turns out that this is impossible because,
as discussed above, these are complementary variables and any such detector
always affects the photons in such a way as to destroy the original +45°
polarisation, so that they emerge at random through both channels of the 45°
analyser.

The above arguments summarise the Copenhagen view of quantum
measurement theory: it is the act of making a measurement that breaks
the quantum superposition and ensures that one or other of the possible
measurement outcomes is actualised. If this procedure is applied consistently,
then the correct answers are obtained in all practical situations, and for
many physicists that is the end of the story. But this approach is subject
to a major objection, the nature of which will be discussed in the rest of
this chapter and the implications of which are the subject of much of the
rest of this book. In practice, we know that the concept of superposition is
applicable to quantum objects such as photons, while collapse occurs when
photons interact with macroscopic objects such as detectors, but the problem
is to devise a clear objective test to decide what is macroscopic. Quantum
physics is the most fundamental theory we know and we might expect it
to be universally applicable. In particular, quantum physics should be able
to explain the properties not only of atomic-scale particles such as single
photons, but also of macroscopic objects such as light waves containing
many photons, or billiard balls, or motor cars, or photon detectors. How
macroscopic does an object have to be before we can expect its state to
collapse?

To explore the question further, we consider again the polarisation measuring experiment, but now modified to include a detector, as in Figure 4.5.
The detector is connected to a meter with a pointer that can be in one of three
possible positions. Position O corresponds to the initial state before a photon
has passed into the apparatus, while positions H and V correspond respectively to the passage of an *h* or a *v* photon through the appropriate channel.

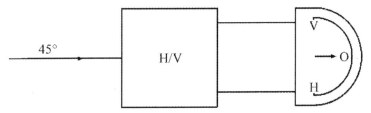

Fig. 4.5 We imagine an H/V apparatus connected to a detector and meter arranged so that the pointer moves from the position O to V if the photon is detected as vertically polarised and to the position H if it is found to be horizontally polarised. However, quantum physics implies that, in the same way as the photon is neither horizontally nor vertically polarised until it has been measured, the pointer is at neither H nor V until a measurement has been made on it.

Now consider what happens when, say, a +45° photon passes through the apparatus. If we treat the system as a measuring apparatus subject to collapse, we expect the pointer to move either to A or to B with equal probability. But now think of the whole set-up as a quantum object. The same argument that told us that the photon is in a superposition of h and v until we record through which channel it passes can be applied to the whole system: the pointer is in a superposition of being in position H and position V *until this position is measured*. Unless this measurement is made, it is always possible to envisage a mechanism similar in effect to the reversed polariser in Figure 4.3. This would recombine the two beams into the original 45° state and restore the pointer to the position O. Only if we make a measurement of the pointer position, e.g. by placing a camera near the apparatus so that it takes a photograph of the pointer before its state is restored, is this possibility removed. But this can only be a temporary solution, because the camera can also presumably be treated as a quantum object whose state is known only when a measurement is made *on it*. This argument can be continued indefinitely and there would seem to be no unique point at which the measurement can actually be said to have occurred.

The central point of the quantum measurement problem can therefore be summarised as follows. Our analysis of the behaviour of microscopic objects like photons shows us that contradictions arise if we attribute properties such as polarisation to them unless these have been measured. However, if quantum physics is to be a universal theory it must be applicable to the measuring apparatus also, which therefore cannot be said to be in any particular state until a measurement has been made on it in turn.

Another way to put the problem is that a 45° photon which has passed through an H/V apparatus without a detector has the potential to behave either as if it has H/V polarisation or, if it is now passed through a reconstructing apparatus, as if it still had 45° polarisation. Only after a measurement has been made is one of these potentialities destroyed. In the same way, if quantum physics is a universal theory then the detection apparatus should retain the potential of being in either of the two pointer positions until a measurement is made on it. The practical problems involved in demonstrating the reconstruction of a detector state, analogous to the reconstruction of the 45° photon state, are immense. Real measuring apparatus and pointers are constructed from huge numbers of atoms and, before such an effect could be demonstrated, these would all have to be returned to precisely the same condition they were in before the photon entered the apparatus, which is completely impossible in practice. Unless, however, it is for some reason impossible *in principle* there is no point at which we can say that the measurement has been made.

Schrödinger's cat

Erwin Schrödinger, who was one of the founders of quantum mechanics (see Chapter 1), graphically illustrated the problems that arise when we consider the effects of measurement on a quantum system. He imagined a situation similar to that set out in Figure 4.6. Inside a large box we have, as well as the familiar light source, polariser and detector, a loaded revolver (or some other lethal device) and a cat! Moreover, the pointer on the detector is connected to the trigger of the loaded revolver in such a way that if a vertically polarised photon is detected then the revolver fires and the cat is killed, whereas a horizontally polarised photon does not affect the revolver and the cat remains alive. When shut, the box containing the cat and apparatus is assumed to be perfectly opaque to light, sound or any other signal that could tell us what is happening inside. We now ask what will happen when a single 45° photon passes through the apparatus. If we regard the cat as a detector the answer is straightforward: the cat is killed if the photon is vertically polarised and remains alive if the polarisation is horizontal. But what does an observer outside the box who understands quantum physics say? Presumably he cannot draw any conclusion about the state of the system until it has been measured which, as far as he is concerned, is when the box has been opened and the state of the cat (dead or alive!) observed. More than this, he will conclude that until this observation has been made a further operation that would restore the photon and the box contents to their original condition is always possible in principle. We therefore cannot say that the state of the system has been changed: the photon is still polarised at 45° to the horizontal and the cat is presumably in a superposition state of suspended animation until it is observed!

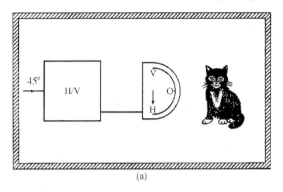

(a)

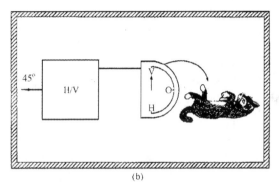

(b)

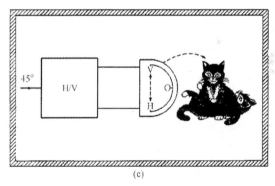

(c)

Fig. 4.6 Schrödinger's cat. (a) If a photon comes out through the horizontal channel of the polariser we expect the cat to be unaffected and remain alive, but (b) if it is vertical a lethal device is triggered and we expect the cat to be killed. (c) Does quantum physics imply that until the box is opened and its state measured the cat is neither alive nor dead?

We can illustrate the point in an even more dramatic way by considering another example of a quantum measurement. It is well known that the evolution of living organisms results from mutation in the DNA of the genetic material of members of a species, which in turn causes a change in the characteristics of the offspring. It is also a fact that such mutations can be caused by the passage of high-energy cosmic ray particles. But these cosmic rays are subject to the laws of quantum physics and each cosmic ray particle has a range of possible paths to follow, only some of which give rise to a mutation. The mutation therefore fulfils the role of a measuring event similar to the detection of a photon. However, if we consider the biological cell as a quantum system, we cannot say whether the mutation has occurred until we make a measurement on it. And if we go so far as to treat the whole planet as a quantum system, we cannot say that the species has evolved or not until we measure this. The world must retain the potential to behave both as if the species had evolved and as if it hadn't, in case a situation arises that brings these two possibilities together and reconstructs the original state, in the same way as the 45° state is reconstructed by the reversed polariser in Figure 4.3!

At this point the sensible reader may well be thinking something like 'Well, if quantum physics is saying that a gun can be half fired and half not fired at a cat who is then half dead and half alive, or that the world contains a biological species that half exists and half doesn't, then this is just ridiculous. I am going to put this book down and forget all this nonsense!' But it is the fact that these implications of quantum measurement theory appear so absurd that is the main point of the argument. However successful quantum physics may have been in explaining the behaviour of atomic and subatomic systems, some of its predictions about detectors, cats and biological systems seem quite wrong. What we hoped would be the final, fundamental theory of the physical universe appears to be flawed. But how is the theory to be modified to become acceptable? At some point in the measurement sequence the chain must be broken so that it is possible to say that a physical system is in a particular state. We have seen that this point is beyond the stage of a single photon and polariser, but whether the change occurs actually in the detector or at a later point is still an open question which is the subject of considerable debate. The Copenhagen interpretation asserts that the change occurs when a detection event happens in a macroscopic apparatus, but there is no clear objective prescription of what is meant by 'macroscopic' in this context.

It may help us to understand this issue better if we consider a completely different piece of twentieth-century physics. As we indicated in Chapter 1, the physics of large-scale objects was well understood in the nineteenth century on the basis of Newton's mechanics and Maxwell's electromagnetism.

However, at the beginning of the twentieth century, Einstein showed that such classical theories had to be replaced by relativistic theories for objects moving close to the speed of light. An important feature of relativity is that it applies to all objects however fast or slowly they are travelling: it is just that the effects are too small to be noticed if the particle speed is much less than that of light. There is a smooth transition between behaviour at low speeds, where relativistic effects are negligible, and high speeds, where they are all-important. The quantum measurement problem could be resolved if there were some parameter in the quantum situation that would ensure a similar smooth transition between a context where super-positions are expected through to one where collapse takes place. However, there is nothing in the standard interpretation of quantum physics to fulfil the role played by particle speed in relativity. We have to either modify the laws of quantum physics as some have attempted (see Chapter 7) or find some other way of looking at this problem.

It is our aim in the rest of this book to discuss some of the ideas put forward in the hope of resolving this issue. We shall be led to consider some strange ideas about the nature of the universe and our place in it. Are we unique creatures with souls and is this essential to any understanding of physical reality? Is there not one universe but are there many, which interact fleetingly during measurements? Or is there some more 'down-to-earth' solution to the problem? We start our discussion in the next chapter with a consideration of the first of these questions.

5 · Is it all in the mind?

In the last chapter we saw how the measurement problem in quantum theory arises when we try to treat the measurement apparatus as a quantum system. We need more apparatus to measure which state the first apparatus is in, and we have a measurement chain that seems to go on indefinitely. There is, however, one place where this apparently infinite sequence certainly seems to end and that is when the information reaches us. We know from experience that when *we* look at the photon detector *we* see that either it has recorded the passage of a photon or it hasn't. When *we* open the box and look at the cat either it is dead or it is alive; *we* never see it in the state of suspended animation that quantum physics alleges it should be in until its state is measured. It might follow, therefore, that human beings should be looked on as the ultimate measuring apparatus. If so, what aspect of human beings is it that gives them this apparently unique quality? It is this question and its implications that form the subject of the present chapter.

Let us examine more closely what goes on when a human being observes the quantum state of a system. We return to the set-up described in the last chapter, where a 45° photon passes through a polarisation analyser that moves a pointer to one of two positions (H or V) depending on whether the photon is horizontally or vertically polarised. At least that is what would happen if the analyser and pointer behaved as a measuring apparatus. If however we treat them as part of the quantum system, the pointer is placed in a superposition of being at H and at V until its state is measured. We now add a human observer who looks at the pointer (it would be possible to imagine the observer using one of his other senses, for example hearing a particular sound caused by a change in the pointer's position, but it is clearer if we think of a visual observation). In physical terms this means that light is scattered from the pointer into the observer's eyes, the retina picks up the signal and transmits it along the optic nerve to the brain. So far the process would seem to be just like that carried out by any other measuring apparatus and there would not seem to be any evidence of a uniquely human act. From now on, however, the measurement becomes part of the observer's knowledge. He is conscious of it. It is in his mind. The attribute of human beings that distinguishes us from other objects in the universe is our consciousness, and if we adopt this approach to the quantum measurement problem, consciousness has an even more central role to play in the physics of the universe than we might ever have imagined.

An example of the distinction between the conscious observer and a more conventional measuring apparatus is illustrated by a variation of the Schrödinger's cat situation known as 'Wigner's friend', after E. P. Wigner, who was central in the development of the consciousness-based theory of measurement. In this example we replace the cat by a human 'friend', and the gun by a conventional detector and pointer. When we open the box, we ask our friend what happened and she will tell us that the pointer moved to H or V at a certain time. Assuming, of course, our friend to be a truthful person, we cannot now treat the whole box and its contents as a quantum system, as the friend would have had to be in a state corresponding to a superposition of knowing that the pointer is at H and knowing that it is at V – until we asked her! The cat may not really have been alive or dead, but the state of our friend's mind is quite certain – at least to her.

A consciousness-based quantum measurement theory therefore relies on the premise that human consciousness behaves quite differently from any other object in the universe. In the rest of this chapter we shall look at some of the evidence for and against this proposition and then try to see whether it can form the basis of a satisfactory quantum measurement theory.

The idea that human consciousness is unique and different from anything else in the universe is of course a very old and widely held belief. Ever since men and women started thinking of their existence (which is probably ever since we were 'self-conscious') many people have thought that their consciousness, sometimes called their 'mind', their 'self' or their 'soul', was something distinct from the physical world. This idea is a central tenet of all the world's major religions, which maintain that consciousness can exist independently of the body and indeed the brain – in some cases in a completely different (perhaps heavenly) existence after the body's death, and in others through a reincarnation into a completely new body, or into the old body when resurrected at a Last Judgment.

A twentieth-century book setting out arguments in favour of the separateness of the soul was written jointly by the famous philosopher Sir Karl Popper and the Nobel-prize winning brain scientist Sir John Eccles in 1977. The book's title *The Self and Its Brain* clearly indicates the viewpoint taken. It is of course impossible to do justice to nearly 600 pages of argument in a few paragraphs, but we can try to summarise the main ideas. Popper starts with a definition of 'reality': something is real if it can affect the behaviour of a large-scale physical object.[1] This is really quite a conservative

---

[1] Popper actually refers to large-scale objects partly to avoid discussing the quantum behaviour of microscopic bodies, so there is a potential problem of consistency here if we are to apply his ideas to the measurement problem.

definition of reality and would be generally accepted by most people, as will become clear by considering a few examples. Thus (large-scale) physical objects themselves must be real because they can interact and affect each other's behaviour. Invisible substances, such as air, are similarly real if only because they exert effects on other recognisably real solid objects. Similarly, gravitational and magnetic fields must be real because their presence causes objects to move: dropped objects fall to the floor, the moon orbits the earth, a compass needle turns to point to the magnetic north pole and so on. Popper describes all such objects, substances and fields as belonging to what he calls 'world 1'. There are two more 'worlds' in Popper's philosophy. World 2 consists of states of the human mind, conscious or unconscious. These must be considered real for exactly the same reason as were world-1 objects – i.e. they can affect the behaviour of physical objects. Thus a particular state of mind can cause the brain to send a message along a nerve that causes the contraction of a muscle and a movement of a hand or leg, which in turn may cause an undoubted world-1 object such as a football to be propelled through the air.

Beyond worlds 1 and 2 is world 3. Following Popper, world 3 is defined as the products of the human mind. These are not physical objects nor are they merely brain states, but are things such as stories, myths, pieces of music, mathematical theorems, scientific theories etc. These are to be considered real for exactly the same reasons as were applied to worlds 1 and 2. Consider for example a piece of music. What is it? It is certainly not the paper and ink used to write out a copy of the score, neither is it the compact disc on which a particular performance is recorded. It isn't even the pattern on the disc or the vibrations in the air when the music is played. None of these world-1 objects *are* the piece of music, but all exist in the form they do *because of* the music. The music is a world-3 object, a product of the human mind, which is to be considered as 'real' because its existence affects the behaviour of large-scale physical objects – the ink and paper of the score, the shape of the grooves in the record, the pattern of vibrations in the air and so on.

Another example of a world-3 object is a mathematical theorem such as 'The only even prime number is 2'. Everyone who knows any mathematics must agree that this statement is true and it follows that it is 'real' if only because world-1 objects such as the arrangement of the ink on the paper of this page would otherwise have been different. The reality of scientific theories is seen in even more dramatic ways. It is because of the truth of our scientific understanding of the operation of semiconductors that microchip-based computers exist in the form they do. Tragically, it was the truth of the scientific theories of nuclear physics that resulted in the development, construction and detonation of a nuclear bomb.

The reader may well have noticed an important aspect of these world-3 objects. Their reality is established only by the intervention of conscious, human, beings. A piece of music or a mathematical theorem results in a particular mental state of a human being (i.e. a world-2 object) which in turn affects the behaviour of world 1. Without human consciousness this interaction would be impossible and the reality of world 3 could not be established. It is this fact that leads Popper and Eccles to extend their argument to the reality of the self-conscious mind itself. Only a self-conscious human being can appreciate the reality of world-3 objects, so it follows that human consciousness itself must be real and different from any physical object, even the brain.

These ideas are developed further in a major section of the book, written by John Eccles. He describes the physiological operation of the brain and speculates on how the self-conscious mind may interact with the brain: he suggests a remarkably mechanistic model in which he postulates that particular 'open synapses' in the brain are directly affected by the (assumed separate) self-conscious mind. An interaction of this kind is a necessary consequence of the idea of a mind or soul that is separate from the body and brain: before an, undoubtedly real, world-1 event can occur there must be an interaction between the 'thoughts' of the mind and the physical states of the brain. At some point there must be changes in the brain that do not result from normal physical causes but which are the result of a literally 'supernatural' interaction.

The above arguments are by no means universally accepted and many people (including the present author) believe that a much more 'natural' understanding of consciousness is possible. But if we accept the idea of our conscious selves as separate from and interacting with our physical brains, a resolution of the measurement problem is immediately suggested. We simply postulate that the laws of quantum physics govern the whole of the physical universe and that the measurement chain is broken when the information reaches a human consciousness. The interaction between mind and matter, which by definition is not subject to the laws of physics, breaks the measurement chain and puts the quantum system into one of its possible states.

The effect of this view of the quantum theory of measurement on our attitude to the physical universe can hardly be exaggerated. Indeed it is difficult to hold this position while still assigning any reality at all to anything outside our consciousness. Every observation we make is equivalent to a quantum measurement of some property that apparently has reality only when its observation is recorded in our minds. If the state of a physical system is uncertain until we have observed it, does it mean anything to say that it even exists outside ourselves? 'Objective reality', the reality of objects outside ourselves, seems, in Heisenberg's phrase, to have 'evaporated' as a result of quantum physics. As Bertrand Russell put it in 1956: 'It has begun to seem that matter, like the Cheshire Cat, is becoming gradually diaphanous and nothing is left but the grin, caused, presumably, by amusement at those who still think it is there'. Of course the existence of the external universe has always been recognised as a problem in philosophy. Because our knowledge of the outside world (if it does exist!) is mediated by our senses, it seems natural to believe that it is this sensual data of whose existence we can be sure. When we say, for example, that there is a table near us, some will insist that all we actually know is that our mind has acquired information by way of our brain and our senses that is consistent with the postulate of a table. Nevertheless, before quantum physics it was always possible to argue that by far the simplest model to explain our sense data is that there really is a table in the room – that the external physical universe does exist objectively. A quantum theory based on consciousness, however, goes further than this: the very existence of an external universe, or at least the particular state it is in, is strongly determined by the fact that conscious minds are observing it.

We have reached a very interesting position. Ever since the beginnings of modern science four or five hundred years ago, scientific thought seems to have moved humankind and consciousness further from the centre of things. More of the universe has become explicable in mechanical, objective terms, and even human beings are becoming understood scientifically by biologists and behavioural scientists. Now we find that physics, previously considered the most objective of the sciences, is reinventing the need for the human soul and putting it right at the centre of our understanding of the universe! However, before accepting such a revolution in attitudes, it is important to examine some of the arguments against a consciousness-based measurement theory and to understand why, although some continue to support it, most physicists do not believe it is an adequate solution to the measurement problem nor, indeed, a correct way to understand the physical universe and our relationship with it.

The main problem with countering a subjective philosophy such as that described above lies in its apparent self-consistency. The basic assumption that the only information we can have about everything outside ourselves is mediated by our senses is incontestable (unless we accept the possibility of *extra*-sensory perception or divine revelation). It seems to follow that it is impossible ever to *prove* the objective existence of the external world. However, there are a number of important arguments that make a purely subjective view, in which the physical world has no objective existence and our consciousness is the only reality, appear at the least unreasonable. Perhaps the most important of these is that different conscious observers agree in their description of external reality. Suppose a number of people driving cars approach a set of traffic lights: if they did not agree with each other about which light was on and what its colour was, a catastrophic accident would certainly result. The fact is that all these drivers experience the same set of sense impressions, which they all attribute to the objective existence of a red light stopping the traffic in one direction, and a green light allowing it to move in the other. Now it is possible to argue that, by some coincidence, their brains and consciousness are all happening to change in similar ways and that the traffic light has no real objective existence, but such an explanation is complex to the point of being perverse compared with the simple objective statement that it really exists.

The extreme, if not the logical, conclusion of subjectivism is to believe that the information received from other conscious beings also has no reality, but is part of one's own sense impressions. Thus my own (or your own?) thoughts are the only reality and everything else is illusion. There is only me: not only the car and traffic lights, but also the other drivers and their states of consciousness are figments of my imagination. Such a viewpoint is known as 'solipsism' and, by its very nature, puts an end to further discussion about the nature of reality or anything else. On the one hand, if everything, including this book and you reading it, are just figments of my imagination there wouldn't seem to be much point in my writing it; on the other hand, if this book and I are just figments of your imagination there wouldn't seem to be much point in your reading it!

We have therefore been led to reject a purely subjectivist philosophy, not because it can be proved inconsistent but because its consequences lead us to statements that, although they cannot be disproved, are complex and unreasonable. A test of simplicity and reasonableness has always formed an important part of scientific theory. It is always possible to invent over-elaborate models to explain a set of observed facts, but the scientist, if not the philosopher, will always accept the simplest theory that is consistent with all the data, because this approach is justified by past experience. There is a (no doubt apocryphal) story about a person who always spread salt on the floor before going to bed at night. The reason for doing so was 'to keep away the tigers'. When told that no one had ever seen a tiger in this part of the world the reply was 'that shows how cleverly they keep out of sight and what a good job the salt is doing'. An important test of any scientific theory is that it should have no 'tigers' – i.e. no unnecessary postulates. The difficulty with theories of quantum measurement is that they all appear to contain 'tigers' of one kind or another and there is no general agreement about which theory contains the greatest number or the fiercest ones! The point of the last few paragraphs has been to show that a theory based on the idea that our subjective consciousness is the only reality is a tiger of the fiercest, or even man-eating, kind!

Just a minute, though! It's all very well to criticise the idea that consciousness is the only reality, but is this really what a consciousness-based theory of quantum measurement is saying? The mind may play a crucial role in the measurement process, even to the point that a choice between possible quantum states is made only when the signal is recorded on a consciousness, but the *existence* of the physical system need not therefore be in doubt. Even if all the properties of the physical system are quantum in nature, in the sense that their values are attained only when a conscious observer observes them, the possible outcomes of these observations are quite outside the control of the observer. The traffic lights can only be red, green or amber – no one can turn them blue or purple just by looking at them. The photon emerging from the H/V polariser is seen to be either horizontally or vertically polarised by any conscious observer: no one can change the polarisation to 45° or double the number of photons passing through the apparatus just by observing it. Is it not possible to maintain the idea of consciousness as the end of the measurement chain without going anywhere like as far as saying that subjective experience is the only reality?

The difficulty with answering yes to the above question is to draw a distinction between the existence of an object, be it photon, measuring apparatus or world-3 concept – and its properties. If all the properties of an object, its mass, position, energy etc., are quantum in nature and do not have values until they are measured, it is hard to see any meaning in the object's separate existence. Over and above this, however, the consciousness-based theory of measurement still leads to some conclusions that are incredible, to say the least, and which correspond to 'tigers' just about as large and ferocious as some of those encountered earlier. A consciousness-based quantum measurement theory states, in brief, that the choice of possible states of a quantum system and its associated measuring apparatus is not made until the information has reached the mind of a conscious observer: the cat is neither alive nor dead until one of us has looked into the box; the species both evolved and didn't evolve until observed by a conscious person. Is it reasonable to think that the presence or absence of a biological species today, and of its fossil record over millions of years were determined the first time a conscious human being appeared on the planet to observe them? Such a view is surely hardly more credible than the suggestion that all reality is subjective.

Such objections have led some thinkers to suggest that consciousness is not just a property of human beings, but is possessed to a greater or lesser extent by other animals (in particular cats!) and even inanimate objects. Alternatively, others have suggested that the world is observed not only by ourselves but by another, eternal, conscious being, whom we might as well call 'God'. The idea that God has a role in ensuring the continual existence of objects that are not being observed by human beings is actually quite an old one and led to the following nineteenth-century limerick:

> There once was a man who said, 'God
> Must think it exceedingly odd
> If he finds that this tree
> Continues to be
> When there's no one about in the quad'

and its reply:

> Dear Sir, your astonishment's odd
> I am always about in the quad
> And that's why the tree
> Will continue to be
> Since observed by, yours faithfully, God

A similar idea has been stated more prosaically by a twentieth-century writer on quantum measurement problems:

> If I get the impression that nature itself makes the decisive choice what possibility to realise, where quantum theory says that more than one outcome is possible, then I am ascribing personality to nature, that is to something that is always everywhere. Omnipresent eternal personality, which is omnipotent in taking the decisions that are left undetermined by physical law, is exactly what in the language of religion is called God.
>
> F. J. Belinfante *Measurements and Time Reversal in Objective Quantum Theory,* Pergamon, 1975.

The difficulty with this point of view is that it restates the problem without solving it. If everything has consciousness, or if God's consciousness is determining which state a quantum system will occupy, then we are still left with the question: at which point in the measuring chain is this choice being exercised? Presumably God doesn't look at the photon passing through the polariser at least until the detector state has changed. Why not? And if not, why not until the information has reached a *human* consciousness. We are simply back where we started, not knowing at what point the measuring chain ends and why. Turning Belinfante's ideas around, the idea of that God chooses is no different, from, and certainly no more satisfactory than, the idea that nature chooses.

This is not of course to say that God cannot exist but only that this idea does not help us to solve the quantum measurement problem. Similarly, although we have seen that a consciousness-based measurement theory leads to unacceptable consequences, we have not thereby disproved the existence of consciousness or the soul. Indeed many people, including some scientists and philosophers, continue to believe in God and in the human soul without caring one way or the other about quantum theory, and most of Popper's arguments for the existence of world-3 objects and of the mind are quite untouched by what we have said so far. Nevertheless, and although it is not strictly relevant to the quantum measurement problem, we shall give a brief outline of some of the modern ideas about consciousness and the brain that argue against the idea of a separate mind or soul.

Let us first say that Popper's arguments, summarised earlier in this chapter, for the reality of world-3 objects and consciousness seem correct and convincing. If a real object is something that can cause a change in a large-scale material object then world-3 objects and consciousness are indeed real and do transmit their effects through world-2 states of the brain. Where many people part company with Popper, and particularly with his co-author John Eccles, is when it is suggested that consciousness is separate from the brain – that there is 'a ghost in the machine'. An alternative view of the link between consciousness and the brain is to draw an analogy with the relationship between a computer program and a computer. A computer is a complex array of electronic switches with no apparent pattern or purpose in itself. It is only when it is programmed – i.e. when the switches are made to operate in a particular sequence – that the computer operates in a useful manner. However, although the program is a world-3 object, it has no effective existence apart from the computer and is certainly not independent of it in a 'soulist' sense. In the same way it is possible that the mind, although real in the same way as the computer program is real, is not separate from the brain. In computer language, the program (consciousness) is 'software' while the computer (brain) is 'hardware'.

This view of consciousness receives some support from research into 'artificial intelligence', in which the capacity of programmed computers to 'think' is investigated. On the one hand, computers can play chess, often holding their own against grand masters, can answer questions in an apparently intelligent way and, when fed with the appropriate information, can recognise human faces. On the other hand, there is considerable doubt over whether the way the computer plays chess, for example, bears much resemblance to the mental processes of a human contestant, and to date computers have failed to excel at games like 'GO', which are believed to need a more intuitive type of thinking. The physicist Roger Penrose has argued that the human mind can carry out tasks that would be impossible, in principle, for any machine following the algorithmic processes underlying a computer calculation.[2] This conclusion is by no means generally accepted, but experts agree that computers are still a long way from behaving anything like a fully conscious human

---

[2] Penrose's ideas are in many ways reminiscent of Popper's arguments for the reality of world 3.

being, able to appreciate world-3 objects and to use this understanding to affect world 1. Even if a computer were built and programmed so that its behaviour was indistinguishable from that of a conscious human mind, some will still argue that the programmed computer is not self-conscious in the same way as a human being is. Nevertheless, the possibility that we will eventually be able to understand consciousness along these lines is now so real that basing a quantum theory or philosophy on its existence as a unique separate non-physical entity must be considered doubtful to say the least.

Quantum physics and reductionism

Modern science is often described as 'reductionist', and only sometimes is this meant as a compliment! The idea of reductionism is illustrated by the fact that I am typing this sentence into a computer. The words appearing on the screen are of course made up of letters, and these in turn are stored in the computer as a set of electronic 'binary bits'. In this way the sentence is 'reduced' to a set of switches that can be 'on' or 'off'. One version of reductionism claims that the only 'reality' is the physical state of these switches and that the meaning of the sentence is in some sense unreal and by implication less important. We could even extend this idea to the thoughts in my mind, if we accept that these are represented by the state of my brain's synapses: only their state would be real and my thoughts and emotions would apparently be devalued. It is the consequences of this form of reductionism that often gives rise to strong opposition. Reductionists are often accused of undervaluing those things many people feel to be most important – ideas, beliefs, law, ethics etc. Indeed, Popper's arguments for the reality of world-3 objects discussed above clearly run contrary to this strong form of reductionism.

However, there is another, perhaps weaker, form of reductionism. The principle underlying it is that our world view should be consistent. By this is meant that it should not lead to inconsistent predictions about the behaviour of physical systems. If we follow the arguments set out at the end of the previous section, we can believe world-3 objects to be of supreme importance, and also that they are manifestations of world-1. This is to say that, although world-3 objects are real and important, when they are encoded on world-1 objects the latter are always subject to standard physical laws. We can perceive the writing on the computer screen to have meaning and importance, while still believing that the computer memory bits obey the laws of physics. The same could be true about my thoughts and their representation by the firing of the synapses in my brain. If this were not the case, the memory bit or synapse would have to behave in a way that is inconsistent with the known laws of physics, as indeed was John Eccles' conclusion in the book discussed earlier. Of course, we may well not know all there is to know about physics and our laws may need amendment or extension. However, a reductionist will insist that this must happen in a manner that is consistent with the fact that the fundamental components of the system (e.g. the atoms in a gas) obey the same laws whatever the macroscopic context.

In practice, physical theories are often devised to explain behaviour in a particular context. For example, the laws of fluid mechanics describe the properties of fluids and are used (not always successfully) to predict quite complex phenomena such as weather patterns. It would be completely impractical to tackle such problems by calculating the motion of every atom in the fluid but we still believe that Newton's laws govern the latter, and the macroscopic laws of fluid dynamics have been developed on the basis that there is no inconsistency between these and a microscopic description. It is a remarkable feature of the physical world that complexity can emerge in a physical system, even though quite simple laws govern the behaviour of its fundamental components. In this sense, reductionism is actually a form of censorship: it tells us what *cannot* happen rather than what can. For example, however complex the weather pattern, no atom in the atmosphere is moving faster than light! Of course, unless and until computers have been developed to be intelligent and conscious in the way discussed earlier, it is a matter of belief that the same principles apply to the relationship between the brain and the mind. Nevertheless, this form of reductionism is the orthodox scientific view, likely to hold until it is disproved experimentally. Indeed, if events occurred that breached this weak reductionism, we would be likely to describe them as 'supernatural' or 'magic'.

The discussion so far has been in the context of classical, deterministic physics, but what about the quantum regime? The whole measurement problem rests on the fact that the laws governing large-scale objects, such as polarisers and detectors, appear not to be consistent with the laws that apply to the quantum behaviour of fundamental entities such as photons. Schrödinger's cat is always seen to be either alive or dead, although quantum physics predicts that it should be in a superposition of these states. It certainly appears as if we have a breach of even the weak form of reductionism: the atoms in the cat change their state in a way that is not predicted by the quantum laws we expect to apply to them. Attempts to resolve the measurement problem can be thought of as ways to avoid abandoning reductionism. Bohr does this by denying the physical reality of unobserved quantum processes, so that there is no lower level of reality to reduce to. The consciousness-based theory of measurement states that the reductionist principle breaks down, but only when the measurement chain reaches a conscious mind. The later chapters in this book describe other attempts to resolve the measurement problem that aim to leave weak reductionism intact.

We referred above to a breach of the reductionist principle being in some sense supernatural or magic, and an example of this would be 'paranormal' phenomena associated with 'extra-sensory perception' (ESP) and so forth. If information passes from one mind to another without any physical mechanism to transmit it, then the synapses in the receiver's brain must have been altered in a way that is not allowed by the physical laws they normally follow. If such phenomena are real (and this is a matter of great controversy) they could be consistent with weak reductionism only if the laws of physics were amended and extended, and it is sometimes suggested that the ideas of quantum physics can be used to achieve this. The implication is that the apparent breakdown of reductionism in the quantum context may be enough to bring these phenomena into the scientific fold. We close this chapter with a short discussion of this idea; we are not going to discuss the evidence for or against the existence of such effects but only whether quantum physics can reasonably be thought to give them any scientific support and respectability.

A typical ESP situation might be where one experimenter sends messages about something like the pattern on a card to another person in a different room, with no known means of communication between them. After this has been repeated a large number of times with many cards, a success rate significantly greater than that allowed by chance is sometimes claimed. From the point of view of quantum physics, we might be struck by an apparent connection with the EPR experiments discussed in Chapter 3, and, indeed, the observed correlations between separated photons have sometimes been described as 'spooky'. Perhaps the minds of the separate experimenters are entangled together in a manner analogous to the connection between the polarisations of widely separate photons. However, such an idea rests on two fallacies. First, an essential part of the experiments discussed in Chapter 3 is the correlated pair of photons emitted from a single source; the quantum predictions (confirmed by the Aspect experiment) are predictions only about pairs of photons created in this way, and there is no obvious equivalent to this in the ESP experiment. Even if we were to be very fanciful and suggest that the minds of the experimenters have in some way been entangled in the past, they could still not send signals to each other because it is impossible to use entangled pairs for signalling (see the discussion towards the end of Chapter 3). We conclude that there is nothing analogous to the photon pair in the ESP case and, even if there were, there would be no way in which quantum physics would allow any communication of information in this situation.

Another way in which it might be suggested that quantum physics could account for ESP is through a consciousness-based measurement theory. If the 45° photon retains its potential to be in a superposition of h and v until it is observed by a self-conscious mind, if the cat is both alive and dead until someone looks at it, then mind is apparently influencing matter. Are we then in a position to explain 'psychokinesis', in which some conscious minds with particular powers are said to be able to cause objects to move around rooms or to bend spoons or whatever? At a less dramatic level, is it possible to understand how (as has been claimed) a conscious observer could influence the time at which a radioactive atom decays? On further examination it again becomes clear that these alleged phenomena are no more consistent with quantum theory than they are with classical physics. Because, even if the mind is the final (or the only) measuring apparatus, it acts *as a measuring apparatus*. It is true that in quantum physics the observed system is influenced by the type of measurement, but this influence is limited to determining the nature of the possible outcomes of the experiment – what we called 'first-level' operations in Chapter 4. The 'second-level' statistical results, such as the number of particles emitted per second by a radioactive atom, are accurately and correctly predicted by quantum theory, independently of the method of observation. Whether or not it is the mind that is finally responsible for the measurement, if a large number of 45° photons have passed through an H/V apparatus, either about 50% of them have emerged in each channel or the laws of quantum physics have been violated.

We again emphasise that we have not attempted to determine the truth or falsity of the existence of extra-sensory perception and related phenomena, but we have shown that we cannot appeal to quantum theory to make them more reasonable or acceptable. Even a consciousness-based quantum measurement theory ascribes a role to the mind quite different from that required in this context. If such phenomena were to be established with the same reliability and reproducibility as is exhibited by, say, the photon pairs in an Aspect experiment, then a further breach of the reductionist principle would be required, and this would have to be beyond our present science, both classical and quantum.

6 · Many worlds

A completely different interpretation of the measurement problem, one which many professional scientists have found attractive if only because of its mathematical elegance, was first suggested by Hugh Everett III in 1957 and is known variously as the 'relative state', 'many-worlds' or 'branching-universe' interpretation. This viewpoint gives no special role to the conscious mind and to this extent the theory is completely objective, but we shall see that many of its other consequences are in their own way just as revolutionary as those discussed in the previous chapter.

The essence of the many-worlds interpretation can be illustrated by again considering the example of the 45° polarised photon approaching the H/V detector. Remember what we demonstrated in Chapters 2 and 4: from the wave point of view a 45° polarised light wave is equivalent to a superposition of a horizontally polarised wave and a vertically polarised wave. If we were able to think purely in terms of waves, the effect of the H/V polariser on the 45° polarised wave would be simply to split the wave into these two components. These would then travel through the H and V channels respectively, half the original intensity being detected in each. In contrast, photons cannot be split, but they can be considered to be in a superposition state until a measurement 'collapses' the system into one or other of its possible outcomes. Up to now, we have argued that collapse must inevitably happen somewhere in the measurement chain – even if only when the information reaches a conscious mind. The many-worlds interpretation of quantum mechanics challenges this assumption and asks what evidence there is for such a collapse. Since the measurement problem discussed in the last two chapters is a consequence of collapse, we can presumably resolve it if collapse can be shown to be unnecessary. To understand this point of view, let us review the arguments that led us to postulate collapse in the first place. First, in an experiment like that illustrated in Figure 6.1 *either* detector D_1 *or* detector D_2 clicks, indicating the arrival of a photon. However, the arguments in Chapter 4 established that if we perform an experiment where the splitting resulting from one polariser is cancelled by another (as in Figure 4.2) then the photon must remain in a superposition state unless and until it is detected. We concluded that, in the absence of collapse, the detector or even Schrödinger's cat would be in a superposition state until *they* were detected. Because this process seemed unending, as well as quite unreasonable and

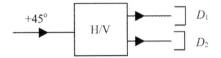

Fig. 6.1 A photon with 45° polarisation passes through an H/V polariser and is then detected in D_1 or D_2.

against all our experience, we postulated the need for collapse and in Chapter 5 examined the possibility that it was the result of human consciousness.

But what actually would be the consequences of assuming that collapse never occurs? We saw in Chapter 4 that the detector has two possible final pointer states, which correspond to the H/V states of the photon. In the same way as the H/V polariser causes the 45° photon to evolve into a superposition of being in one channel *and* in the other, the photon's entry into a measuring apparatus should result in this also being placed in a superposition state. To demonstrate this superposition state in the case of the photon we considered the action of a reverse polariser, so in order to demonstrate that the detector was also in a superposition state we would need to reverse its action in some analogous way. However, this is completely impossible in any conceivable practical situation. The reversal of the photon's polarisation requires quite careful experimentation, so that the two waves come together in step. In order to perform a similar action on a detector or cat, the quantum waves associated with each particle in it would have to come into step at exactly the same point. As there are a huge number (say 10^{20}) of particles in such an object, the chances of this happening are incredibly small (somewhere around 1 in 10 to the power of 10^{20}). Without a reversal experiment the two parts of the photon superposition cannot affect each other, so if the equivalent process for the detector is impossible in practice then, in turn, the two parts of its superposition can never affect each other and there can be no direct evidence that collapse has occurred. Nevertheless, if the argument is correct, the superposition is still real.

What about the arguments at the beginning of Chapter 5 that were posited on the fact that the conscious human observer sees only one outcome? Surely, as we argued, collapse must occur at this point if nowhere else. However, just suppose that it does not. Then, if a human observer also obeys the laws of quantum physics, he or she must also be placed in a superposition state. The arguments showing that the two halves of such a superposition cannot affect each other now mean that the two parts of an observer in a superposition state are forever unaware of each other's existence. In other words, even if there is a human observer in the chain, he or she is split into two: one half of this superposition believes that the photon has emerged in (say) the H and not the V channel and the other believes the opposite. 'Now, this is really ridiculous', I hear you say, 'If measuring apparatuses, not to mention people, were splitting and multiplying in this way I would certainly have heard about it, or noticed it when it happened to me!' However, the whole burden of the above argument is to show that you would not. This is the absolutely crucial point about the many-worlds interpretation: *once a split has occurred between states involving macroscopic objects, such as detectors, the two branches have no practical way of affecting or being aware of each other.* Thus when the 45° photon is split in the H/V polariser and interacts with a measuring apparatus there is no interaction between the components of the resulting superposition, and the same is true when a person divides. The two copies of the measuring apparatus or people continue to exist, but they are completely unaware of each other's existence. It follows that because each measuring apparatus is interacting with the world about it and indeed eventually with the rest of the universe, each component of the split observer must in fact carry along with her a copy of the entire universe. Hence the term 'branching universe': whenever a quantum measurement occurs the universe branches into as many components as there are possible results of the measurement. Everyone in a particular branch thinks that their result of the measurement and their particular universe are the only ones that exist (unless of course they know about and believe in the many-worlds interpretation of quantum physics!). In practice, the different branches of the universe never come together again.

We might imagine that we could test the many-worlds hypothesis in principle by performing an experiment in which one of us was split by a quantum measurement and reunited again. The person who underwent this experience could then tell us what it felt like to split and could remember having observed the photon in each of its polarisation states. The problem with this idea is that the recombination of the two halves is possible only if at the same time all information about which channel was followed is erased in the same way as the undetected photon has its original 45° polarisation restored by the reversed calcite crystal. It follows that if a human observer is involved then any memory of the outcome of the experiment must also be erased. Indeed it used to be generally believed that reunification would be impossible unless the observer were restored to the same state as before the experiment, so that any memory of the split would also be erased. Theoretical work by Professor David Deutsch indicates that this may not be correct and that if such an experiment could be carried out, an observer could tell us what the split felt like, but not what the result of the experiment was. He has suggested that if a sufficiently intelligent computer were built sometime in the future, it could play the role of the observer in such a process. If, indeed, human consciousness is different from a computer in not being subject to many-worlds splitting, this could conceivably be tested experimentally by comparing human and computer experience in such a situation.

The postulates of the many-world interpretation may be incredibly radical, but this is at least partly compensated by the resolution of many of the puzzles and paradoxes of the measurement problem. Consider the theory applied to Schrödinger's cat. Instead of having to worry about how a cat can be neither dead or alive, we simply have a superposition of two cat states, one alive and one dead, each in its own box in its own universe; and two observers each opening their own box and drawing their own conclusions about the state of the cat and hence of the photon. Or think of Wigner's friend. We no longer have to worry about whether the measurement is made when the friend observes the photon or when Wigner learns the result: as the photon passes through the measuring apparatus into a detector, it is split and so are Wigner, friend and all. In one universe the friend sees a horizontally polarised photon and tells Wigner this result. In another she sees a vertically polarised photon and passes this message on.

This is why its supporters believe that we should take this theory seriously despite its ontological extravagance reflected in the multiplicity of universes. The solution of the measurement problem combined with the fact that no collapse postulate is needed means that many worlds is actually a simpler theory than its competitors. Reductionism (discussed near the end of Chapter 5) is maintained because a single theory applies to everything from photon polarisation to measuring apparatus and conscious observers. The physicist Paul Davies has described it as 'cheap on assumptions, but expensive on universes!'

Before trying to decide whether the net balance of this account is profit or loss, let us look at a few more consequences of the many-worlds assumptions. We might wonder how many universes there are. As every quantum event has two or more possible outcomes and as huge numbers of branching events are continually occurring, the number must be quite immense. It has been estimated that there are about 10^{80} elementary particles in the observable part of any one universe. If we assumed that each of these had been involved in a two-way branching event each second in the 10^{10} years since the 'big bang', the number of universes created by the present time would be something like 10 to the power of 10^{12} – an unimaginably huge number, and this is certainly a lower estimate. Where are all these universes? The answer apparently is that they are all 'here' where 'our' universe is: by definition universes on different branches are unable to interact with each other in any way (unless they are able to merge in the very special circumstances mentioned earlier) so there is no reason why they should not occupy the same space.

The possibility of the existence of other universes, some of which differ only slightly from our own, can lead to enticing speculations. Thus there would be many universes which contain a planet just like our own earth, but in which a particular living species is absent, because the quantum events that caused the mutation which led to the development of the species in our universe (and many like it) did not follow this path in these other universes. Presumably there are universes in which life evolved on Earth, but humankind didn't, and which are therefore pollution free and not threatened by ecological disaster or nuclear extinction. If we consider that every choice between possible outcomes is fundamentally a result of a quantum event, then every such possibility must exist in its own universe. Every choice we have made in our lives may be associated with a quantum event in our brains and, if so, there would be universes with other versions of ourselves acting out the consequence of all these alternative thoughts. Some of the other universes will be delightful places: because of a happy coincidence of quantum events your favourite football team wins every trophy every year and society is constructed the way you would like it because your favourite political party wins every election. In others, of course, the opposite happens and in many the possibility of irrational political behaviour resulting in the extinction of mankind in a nuclear confrontation or ecological disaster must already have been realised.

Despite all we have said, it would be surprising if many readers have been convinced of the plausibility of many-worlds theory. In the vast majority of the universes that have evolved since you began reading this chapter, you are completely sceptical about the whole idea! To postulate the existence of a near-infinite number of complete universes to resolve a subtle theoretical point seems to be introducing 'tigers' into our thinking with a vengeance. We have seen that the other universes do not interact with us except possibly in very special circumstances when no information is obtained about them in any case. In which case, does the statement that they exist mean anything? In his original paper on the many-worlds interpretation Everett compares such criticisms with the objections of the mediaeval world to Copernicus's suggestion that the earth revolves around the sun. In the same way that people were prejudiced by their entrenched belief that the earth is the centre of the universe, it is suggested that we find the branching-universe idea unaccept-able simply because we have been brought up with the prejudiced idea that only one universe exists. He would say that in both cases acceptance of the radical postulate allows us to make sense of experimental results that were previously illogical and confused. Nevertheless, most scientists would not agree that the two cases are parallel and would maintain that the ontological extravagance of the original postulate is much greater in the many-worlds case and the gain in clarity of understanding is much less. To adapt Paul Davies's words, most would believe that the losses involved in the 'extravagance with universes' heavily outweigh the gains from the 'economy with postulates'.

Over and above the question of ontological extravagance, there are other reasons for criticising the many-worlds interpretation. In particular, there is the problem of defining probabilities in a context where, instead of alternative occurrences, everything possible happens. To explore this point more closely, consider four separate sets of apparatus (including detectors) for measuring H/V polarisation. We allow a photon to pass through each of the sets of apparatus and we assume, as usual, that each photon is in a state of 45° polarisation before the measurement. Consider what the many-worlds model has to say about this sequence of operations. Initially there is only one universe in which none of the four detectors has registered a photon: this is described by the symbol OOOO at the left of the network in Figure 6.2. Suppose now that one of the photons passes through one of the polarisers; this has the effect of splitting the universe into two, one of which has its detectors in the state OOOH and the other in the state OOOV. We continue to pass the photons one at a time through the other sets of apparatus until all four have been measured. We now have a total of 16 universes representing all the possible permutations of H and V, as shown on the right of Figure 6.2. If we now examine how many times H and V photons were recorded in a particular universe we find that six universes recorded 2H and 2V, four recorded 1H + 3V, another four 3H + 1V and in only one universe each were 4H + 0V or 0H + 4V observed. The standard analysis of this experiment as described in Chapter 2 predicts that a 45° photon passing through an H/V polariser will have a 50% chance of emerging through either channel, and this result is confirmed in the majority of universes considered in our example. With only four photons, the fraction of universes which deviate from the ideal 50–50 behaviour is quite large, so that an observer will not be very surprised to observe 3:1 or even 4:0. If the number of photons-plus-apparatuses is increased then the distribution will get nearer and nearer to the ideal 50–50 division, and this is also found to happen in the branching-universe model. For example, a network similar to that shown in Figure 6.2 but with ten photons and ten polarisers produces 1024 universes, 622 of which are within 10% of the ideal 50–50 distribution while the states 10H and 10V have only one universe each. By the time we have hundreds or thousands of photons, the discrepancies from the predicted distribution become extremely small.

At first sight, then, many-worlds theory seems to be in complete agreement with what we expect about the probabilities of different measurement outcomes. The only reason it comes out so simply, however, is that we have

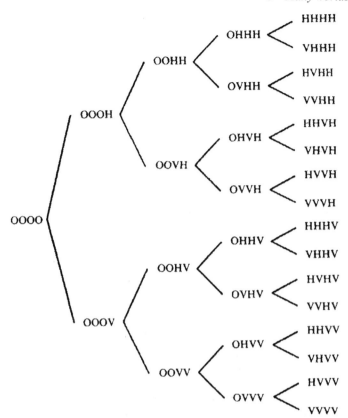

Fig. 6.2 We consider, from the many-worlds point of view, four separate measurements of the H/V polarisation of 45° photons. Beforehand, all four measuring apparatuses are in their initial state O. After the first measurement the universe splits into two branches: in one branch a photon and measuring apparatus are in the state H, while in the other they are in the state V. Further splitting occurs at each subsequent measurement and, when the process is completed, the results are approximately equally divided between H and V in most, but not all, universes.

considered the particular case where the predicted probabilities for the possible outcomes (H or V) are equal. Consider a more general example: the measurement of H/V polarisation on a photon initially polarised at 30° to the horizontal. From the discussion in Chapter 4, the predicted probabilities of obtaining H or V in a measurement are now 75% and 25% respectively. However, if we re-examine the above argument and the many-worlds picture illustrated in Figure 6.2, we can see that these are completely unaffected by the change in the initial state of the photons. The number of universes with a particular pattern of polarisations is determined only by the number of possible results of the measurement. Hence, even in the 30° case, after a large number of measurements are made, the vast majority of them have equal numbers of H and V outcomes. Ever since the many-worlds model was suggested, a major challenge to its supporters has been to reconcile its predictions with the probabilities calculated from standard quantum physics and confirmed by experiment.

Discussions of probability can be quite frustrating because of the difficulty of defining it in the first place. For example, when we toss a coin, what do we mean when we say that there is a 50–50 chance that it will come down heads? One answer is that whenever we have performed an experiment such as tossing the coin 1000 times, it has come down heads about 500 times. But suppose it hadn't; suppose it had come down heads 750 times and tails 250 times. There are then two conclusions we could draw: either it was just a happenstance (because even with a fair coin, there is a small chance of obtaining a sequence with three times as many heads as tails) or the coin was loaded or biased in some way. We could address this question by examining the coin and testing for bias and, if we found a physical reason for expecting it to come down heads rather than tails, we might be able to predict that the chances of this were around 3:1, which would be in line with our observation. Another example is a flowing river or stream, which branches in two, with three times as much water emerging in one channel as the other (Figure 6.3). If objects that float (e.g. leaves or twigs) are dropped into the water at random points upstream, about three-quarters of them will emerge in one channel and one quarter in the other.

Suppose we postulate that the photon measurement experiment is similarly biased. Calculating probabilities by just counting universes assumes that the experiment is unbiased, as in the case of tossing a fair coin. Measurement of H/V polarisation on a photon initially polarised at 30° to the horizontal could be thought to be more akin to the tossing of a biased coin or the dropping of leaves into the branching stream. By assigning a bias (often called a 'weight' or a 'measure') to the outcome, we might be able to restore the probability concept. However, there is a vitally important difference between these classical examples and many-worlds quantum physics. In the classical case, the coin comes down *either* heads *or* tails, but in the quantum case, the photon remains as *h* in one universe *and v* in another. It is as if in the branching stream case there were no twigs or leaves, but just water. More water comes down the broad channel, but it is not meaningful to say that water is 'more likely' to be found there, as it is actually in both. It is arguable that the whole concept of probability is posited on a disjunction – a sentence containing the word 'or' – and is devoid of meaning in the context of a conjunction, i.e. 'and'. In attempting to counter this, supporters of the many-worlds theory emphasise that no observer is able to stand outside the system and judge the relative probabilities by counting universes. What we call probability is actually what is experienced by an observer who is part of the quantum system, splits and all. It is the experience of such observers that determines what we (possibly wrongly) call probability in this context. After undergoing a large number of measurements with their accompanying splitting, the observer can use this experience to make an informed judgement about the future. At each measurement the observer splits, but in any universe an observer will have a memory of a history of measurements and she can use these to form a judgement of what she would call a probability for the possible outcomes of a subsequent measurements. For example if asked to bet on the outcome of an H/V measurement on a 30° photon, she would accept odds of up to 3:1 against it emerging in the vertical channel. In one universe she would have won the bet and in another universe she would have lost it. She is likely to be quite impervious to arguments purporting to show her conclusion to be meaningless if these are based on the existence of other universes of which she can never be aware. The original name for many-worlds theory was 'the relative state interpretation', which can be interpreted as meaning that probability is *relative to* the branch occupied by the observer. If we were able to look on the whole system from outside, we would be right to say that this apparent probability is an illusion, but to a real observer it remains a reality. Nevertheless, there is no doubt that, across all the universes, there will be a larger number of copies of the observer whose experiences do not match the probabilities predicted by quantum physics than there are who do. If we do accept that probability can be usefully defined relative to an observer, the question of whether the size of this probability has to be added to quantum physics as an additional postulate or whether it can be deduced from the other postulates of quantum physics remains a matter of considerable controversy.

We should note the reference to the 'observer' in the above discussion. As we have seen, to resolve some of the problems of this approach we have had to consider carefully what a (presumably human) observer would experience. Much work in the field during the last years of the twentieth century has been concentrated on the consequences of many-worlds theory for our understanding of brain, mind and consciousness. Some workers have tried to develop a 'many minds' theory in which the branching begins (though it cannot end) in the human consciousness. Others have concentrated on a quantum analysis of the behaviour of the brain. However, most work in this area eschews the idea that the observer acts outside the laws of physics, as was suggested in Chapter 5.

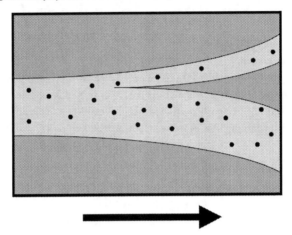

Fig. 6.3 Water flowing in a stream divides into two channels. If we drop floating objects into the stream, we may be able to say that they are more likely to emerge in the wider channel. However, it is meaningless to say that water is more probably in one channel then in the other. Similarly, if a quantum system is in a superposition of two states, it may be meaningless to say that there is a probability of it being in either one or the other.

A belief in many-worlds theory can profoundly affect one's approach to quite wide philosophical issues, even ethics. Is a moral choice meaningful if both consequences are going to occur in one universe or another. An illustration of the type of question that arises is a quantum version of Russian roulette. Imagine a 30° photon approaching an H/V apparatus and suppose the H outcome to be connected to a gun that will fire when the photon is detected, so that if I stand in front of the gun there is then a 3 in 4 chance that I will be killed. I then place an even-money bet with a bookmaker that I will stay alive, and this is accepted because the odds are in the bookmaker's favour. I now carry out the experiment. In one branch I will be dead and the bookmaker will have his money, but in the other, I will be alive and rich! A willingness to carry out such an experiment might well be a test of faith for a believer in the many-worlds interpretation of quantum physics!

A rather more technical question is defining what components form the basis of the branching and when the branching occurs. We saw that the measurement of the H/V polarisation of a 45° photon leads to the apparatus being put into a superposition state of having recorded H and also V. We then said that this superposition branches in an analogous way to the river illustrated in Figure 6.3. But this is a quantum system, not a river. Why can we not define some other superposition (e.g. 30° and 120°) as a basis for the branching? Are we in danger of re-introducing subjective judgement into the theory – just what an objective theory such as many worlds is supposed to be eliminating? For some time, indeed, many-worlds theory was criticised on the grounds that branching is an additional postulate, much as collapse is in the Copenhagen interpretation. This view was encouraged by statements from some of the supporters of many worlds, who talked about 'splitting' rather than branching. However, if this were the case then the many-worlds view has signally failed to solve the measurement problem! The measurement problem arises because of the need to introduce collapse as an additional postulate, so if all we have done is replace this with splitting, the 'economy with postulates' is beginning to look pretty thin. However, modern work has emphasised the idea, implicit in Everett's original formulation, that applying the standard quantum rules to the irreversible changes associated with measurement, or 'measurement-like' processes, results in the system's branching into a basis defined by the possible states of the macroscopic apparatus. In Chapter 5, we quoted Bohr's reference to the settings of the apparatus (e.g. the polariser orientations) as influencing *'the very conditions that define the possible types of prediction regarding the future behaviour of the system'*. In the many-worlds context, these same settings define the basis into which the whole system, including the apparatus and observer, will inevitably branch if the basic rules of quantum physics are followed. We return to this point in Chapter 10.

It is interesting to consider the application of the many-worlds model to the EPR situation discussed in Chapter 3. There we discussed the results of measuring the polarisations of pairs of photons emitted by a common source in a state where their polarisations are at right angles to each other. Referring back to Figure 3.2, we see that whenever one photon is observed in (say) the horizontal channel of an H/V apparatus, its partner behaves from then on as if it were vertically polarised. Before we can analyse this from the many-worlds point of view, we have to know more about how the initial state of the pair is described by quantum physics. Similarly to the way in which the 45° state of a single photon can be thought of as a superposition of h and v states, the initial state of a correlated photon pair can also be described as a superposition of two states. One state represents an h photon on the left and a v photon on the right, while in the other these polarisations are interchanged. Consider now the measurement of H/V polarisation on, say, the left-hand photon. From a conventional point of view, the state of the pair collapses into one or other of the components of the superposition: we observe *either* an h photon in the H channel of the left-hand polariser along with a distant v photon, *or* a v photon in the V channel of the left-hand polariser accompanied by a right-hand h photon. In Chapter 3, we saw how difficult it is to reconcile this behaviour with the idea of locality. We now consider this process from a many-worlds viewpoint, where, as we have seen, there is no collapse: the left-hand photon causes the left-hand apparatus and observer to branch into a superposition. Each branch contains a right-hand photon, but the superposition has not collapsed. It appears as if the state of the right-hand photon has been unaffected by the left-hand measurement, which has interacted only with the photon in its own locality. However, suppose our left-hand polariser had been oriented to measure $\pm45°$ instead of H/V. We would then have expressed the initial state as a superposition of a state with $+45°$ on the left and $-45°$ on the right and another with these polarisations interchanged, and indeed we are free to do so. Identical arguments to the above would then lead us to conclude that the right-hand photon is $+45°$ in one branch and $-45°$ in the other, following a left-hand measurement. There is clearly a correlation between the type of measurement made on the left and the state of the right-hand photon in each branch. Whether this means that many-worlds theory avoids the problem of non-locality is still a matter of controversy.

Even if the many-worlds postulate does resolve the problem of non-locality, the question of probabilities remains a significant difficulty, so it is hard to claim that many-worlds theory provides a complete solution to the measurement problem. What the interpretation does do is recover some kind of realistic description of quantum objects. We can preserve the simple model in which a photon follows a particular path and is detected with a particular polarisation at the same time as keeping a record (in another universe) of what would happen if another path had been followed. Even a form of determinism is recovered because all possible outcomes of a quantum event not only can, but also do, occur. We mentioned in Chapter 1 how Laplace in the nineteenth century summed up the principle of determinism in the statement that 'the present state of the universe is the result of its past and the cause of its future'. In a similar way, the many-worlds physicist believes that the state of the particular universe we happen to be in at a given time is 'the result of its past' and is the 'cause of the future' of all the universes that will branch out from this one. In principle (but not of course in practice) the future states of all these universes could be calculated from the quantum laws and the present state of our own branch. However, it is interesting to note that it is impossible, even in principle, to make the same calculation about the past, as this would require knowledge of the state of presently existing branches other than the one we happen to occupy.

Of course if we reject many worlds, we are back with the measurement problem. If the preservation of all possible states in a multiplicity of universes is unacceptable then a choice of one out of the possible quantum paths must be made somewhere. And if it is also unacceptable that this choice be made only in the conscious mind then it must be made somewhere else in the measuring chain. Is there a point in the process, outside the human mind, at which we can say that the chain is broken and the measurement is complete? The rest of this book explores this question further.

7 · Is it a matter of size?

The last two chapters have described two extreme views of the quantum measurement problem. On the one hand it was suggested that the laws of quantum physics are valid for all physical systems, but break down in the assumed non-physical conscious mind. On the other hand the many-worlds approach assumes that the laws of physics apply universally and that a branching of the universe occurs at every measurement or measurement-like event. However, although in one sense these represent opposite extremes, what both approaches have in common is a desire to preserve quantum theory as the one fundamental universal theory of the physical universe, able to explain equally well the properties of atoms and subatomic particles on the one hand and detectors, counters and cats on the other. In this chapter we explore an alternative possibility, that quantum theory may have to be modified before it can explain the behaviour of large-scale macroscopic objects as well as microscopic systems. We will require that any such modification preserves the principle of weak reductionism discussed towards the end of Chapter 5.

The first point to be made is that the problems we have been discussing seem to make very little difference in practice. As we emphasised in Chapter 1, quantum physics has been probably the most successful theory of modern science. Wherever it can be tested, be it in the exotic behaviour of fundamental particles or the operation of the silicon chip, quantum predictions have always been in complete agreement with experimental results. How therefore have working physicists resolved the measuring problem? The answer is quite simply that practical physicists apply the Copenhagen interpretation, knowing perfectly well when a measurement has been made and what the distinction is between a quantum system and a measuring apparatus. The quantum state of a photon detector, with its photomultiplier tube and high-voltage supply, could never in practice be restored after a measurement to what it was before. Physicists therefore have no practical difficulty in distinguishing between 'measurement' processes, in which for example a particle is detected by such an apparatus, and 'pure quantum' processes, like the passage of a photon through a polariser without detection. Nevertheless, if asked to define this distinction in a consistent way, most of us would find it hard to say clearly where the dividing line comes or why. If we are to resolve the measurement problem in this way, we will have to find a means of making the distinction in a consistent objective manner.

The first parameter that comes to mind in this context is size. Quantum physics applies to atomic and sub-atomic systems, but not to large-scale macroscopic apparatus. However, by size we can't just mean the physical dimensions of the measuring apparatus, because in a typical interference experiment the photons travel many centimetres between the source and the screen, and the interference pattern still appears. Moreover, we saw in our discussion of the EPR problem in Chapter 3 that quantum correlations between pairs of particles can be observed over large distances (several metres in the Aspect experiment and several kilometres in later work). What other property, then, might mark the distinction between a measuring apparatus and a quantum system on which measurements are made? In the present chapter we explore the possibility that the key factor may be the fact that the measuring apparatus consists of a large number of atomic particles.

An obvious feature distinguishing measuring instruments such as photo-multiplier tubes, counters, cats and human beings from pure quantum systems is that the former contain huge numbers of atoms that themselves consist of electrons, protons and neutrons while the latter typically consist of a single particle such as a photon or, at most, a small number of such particles. More-over we can explain the behaviour of measuring instruments, certainly at the point where the information is processed to a stage where we can read it, in largely classical terms. Thus pointers in meters move across dials because electric currents are passing through coils, which then are affected by mag-netic fields; or lights are switched on and the emitted light is intense enough to be treated as a classical wave, ignoring its photon properties. Could it sim-ply be that objects containing a sufficiently large number of particles obey the laws of classical physics rather than quantum mechanics? Microscopic bodies would then obey quantum laws until they interacted with large bodies when a measurement interaction would take place.

One difficulty in this approach is to define just what is meant by 'large enough' or 'sufficiently many atoms' in this context. However, a typical measuring instrument might contain at least 10^{20} atoms which is an awful lot more than one or two. It is conceivable therefore that the truly fundamental laws of physics contain terms of which we are presently unaware and which have a negligible effect when there are only a few particles in a system but become appreciable when the number of particles is large. These might conceivably lead to the collapse of the superposition as soon as such a macroscopic object was involved in the measurement chain. The term for such postulated behaviour is 'spontaneous collapse' and thinking along these lines has led to a reappraisal of the evidence for the quantum behaviour of macroscopic systems.

It has long been realised that, although macroscopic bodies may appear to behave classically, many of their properties in fact result from the quantum

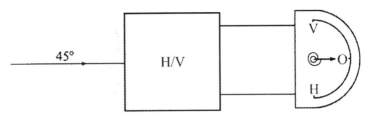

Fig. 7.1 If the pointer on a photon detector is mounted on a perfect spring, will it stand still because the motions induced by the horizontal and vertical signals cancel each other? Or does the pointer always move to either H or V because quantum physics cannot be applied to such a macroscopic object?

behaviour of the system's microscopic components. For example, the fact that some substances are metals, which conduct electricity, while others are insulators or semiconductors, results from the quantum properties of the electrons in the solid. Similarly, the thermal properties of a solid, such as its specific heat or its thermal conductivity, depend on the quantum motion of the atoms in it. It might be thought therefore that the quantum behaviour of macroscopic systems is a well-established fact. However, there is one feature of the measurement process that is not reflected in such considerations. This is that when a record is made of a measurement, the states of a large number of particles are changed together. For example, when a pointer moves across a dial all the atoms move together; when a light is switched on it is the collective movement of the electrons in the wire that gives rise to the electric current that lights the bulb. Such co-operative motion of a large number of particles is an essential feature of all observable macroscopic events and of measuring instruments in particular. Could it be that quantum theory breaks down just when such large-scale collective motions are involved? To test the applicability of quantum physics in such a situation we might consider using an apparatus like that illustrated in Figure 7.1: here we have the usual example of a +45° photon entering an H/V apparatus, the output of which is connected to a detector that swings from the central position O to the points H or V, depending on through which channel the photon emerges. To this extent the apparatus is just the same as that discussed several times before (see Figure 4.5) but we now include a crucial modification. As shown in Figure 7.1, we imagine that the pointer is mounted on a spring so that, after being deflected to one side, it returns to the centre and swings across to the other side, and goes on swinging in this way like a pendulum. It follows that if the photon certainly emerges through, say, the H channel (perhaps because the V channel is blocked) the pointer will swing first to the H side, then to the V and so on, while it will swing in the opposite sense if the photon emerges through

V. If, however, the photon passes through both channels and if quantum theory is applicable to both the photon and the macroscopic pointer, then the pointer will actually follow the sum of both these motions, which in fact cancel each other out leaving it as it was.[1] If we were to perform such an experiment and observe no motion of the pointer, we would have verified the applicability of quantum theory to such large-scale systems. In contrast, if the measurement chain is broken whenever a many-atom measuring apparatus is involved, *spontaneous collapse* would occur, the pointer would always swing in one direction or the other when a photon passed through and the quantum cancellation would never be observed.

A more detailed spontaneous collapse theory was developed in the 1980s and is known as GRW theory after the three Italian scientists[2] who devised it. They started from the postulate that although atomic particles can be in delocalised quantum states that are superpositions of states corresponding to different positions in space, there is a non-zero probability that these will spontaneously collapse into states that are localised close to one position or another. To maintain consistency with the experimental evidence for the quantum properties of atoms and subatomic particles, they postulated that this probability is very small – so small indeed that the state of a particular particle is expected to collapse only once in 10^5 years. However, a macroscopic object like a pointer in a measuring apparatus contains about 10^{20} atoms, so that the state of one of these atoms is likely to collapse every 10^{-8} seconds. GRW pointed out that if one atom in a delocalised pointer were to collapse on its own, it would have to be torn out of one component of the superposition and given double weight in the other. This would inevitably result in an increase of the energy of the system and there is nowhere for this energy to come from. To avoid this, they postulated that the collapse of a single particle in a delocalised macroscopic object inevitably leads to the collapse of the whole object. A pointer therefore remains in a delocalised superposition for only a very short time before collapsing into one of the states corresponding to a measurement outcome. The details of the theory imply that the probability of collapse into any particular state is the same as that predicted by the rules of quantum measurement. If GRW theory were true then spontaneous collapse would solve the measurement problem while preserving weak reductionism, because it is a single theory whose manifestations differ in the macroscopic and microscopic contexts.

The next step is to see whether there is any experimental evidence to support these ideas. Unfortunately, an experiment such as the observation

---

[1] This actual experiment is not practically possible, for reasons to be discussed later.
[2] G. C. Ghiradi, A. Rimini and T. Weber.

of the oscillating counter discussed above is impossible to perform on a conventional counter or measuring apparatus because, even if there were no collapse, the stationary state at O would be reconstructed only if the two oscillations cancelled each other out completely. The pointer is a macroscopic object composed of a very large number of atoms, *all* of which would have to move in exactly the opposite way in the two modes of oscillation. As well as the collective motion associated with the measurement, the position of the pointer is subject to random fluctuations due to its interaction with the air it is passing through and to variations in the frictional forces in the pivot. Even if these could be eliminated, the atoms in the pointer would still possess their more or less random thermal motion, and as a result there is no real possibility of observing a cancellation of the two oscillations. Indeed, more detailed calculations show that, because of this, the pointer would always in practice be found in one oscillation state or the other even if pure quantum behaviour were possible in principle. Alternatively, from a many-worlds point of view, the thermal motion causes a branching that results in one copy of an observer seeing the pointer swing one way while the opposite is observed in another universe. Thus, experimental complications would make the observation of a macroscopic object in a superposition state very difficult even in the absence of spontaneous collapse. As an alternative, some scientists have suggested that the presence of such random thermal motion is an essential part of the measuring process and we discuss these ideas further in the next chapter. For the moment, however, we are interested in the question of whether pure quantum behaviour of a macroscopic body (such as was suggested for the pendulum in the above example) can ever be observed in principle. This possibility has been the subject of considerable experimental investigation in recent years.

The possibility of spontaneous collapse might also be tested by an inter-ference experiment similar in principle to that used to demonstrate two-slit interference for light (cf. Figure 1.2). The wave used would have to be the matter wave associated with a macroscopic object. If all the conditions were such that quantum physics predicted the formation of an interference pattern and this was not observed, we might well conclude that spontaneous collapse had occurred. A large object whose quantum properties have been directly studied in this way is the Buckminster-fullerene molecule. This consists of 60 carbon atoms arranged in a nearly spherical configuration, reminiscent of a soccer ball; for this reason, these molecules are sometimes called 'bucky-balls'. In 1999, a group of scientists working in Vienna and headed by Anton Zeilinger performed an experiment in which buckyballs produced by a fur-nace were passed one at a time through a grating containing a number of slits. (Although there were many more slits than the two discussed in Chapter 1, the principles are exactly the same.) The emerging beam was detected and found to form an interference pattern that is completely explicable on the basis that the beam has wave properties, so that the particles cannot be thought of as passing through any particular slit and no spontaneous collapse has occurred. There are a number of reasons why such an experiment is much harder to perform with molecules than with photons. First, the molecule's large mass means that it has quite large momentum even when moving slowly, so the predicted wavelength of the associated de Broglie wave is very small indeed (around a billionth of a metre in the present case). For waves of these wavelengths to form a visible interference pattern, the slits used also have to be extremely small. Second, C_{60} is a large molecule at a high temperature, so its component atoms are in continual motion and it is an essential requirement that these motions are slow enough that the state of the molecule is virtually unchanged while it is undergoing interference, i.e. that its state remains coherent. Any process that would in principle allow an observer to determine through which slit the molecule passed would destroy this coherence. In fact it is estimated that at the temperature of the experiment, C_{60} would be expected to emit two or three infrared photons during its passage through the apparatus. However, the wavelength of this radiation is much greater than the distance between neighbouring slits, which means that they carry no information about which slit the molecule passed through and therefore coherence is essentially preserved. All these factors are expected to become more important the larger the objects studied.

We can conclude that the successful observation of quantum coherence in this case means that spontaneous collapse does not occur for an object containing about 100 atoms, which is about ten times as many as in any previous experiment. However, a buckyball is hardly a macroscopic object (compared with, say, a small grain of sand containing 10^{12} atoms) and GRW theory predicts that we would have to wait 100 years to observe a collapse of the buckyball state. There are at least ten orders of magnitude to go before we can draw any conclusions about the spontaneous collapse of macroscopic objects.

Superconductivity

There is one class of phenomena that is known to exemplify macroscopic quantum effects in a very dramatic way. This comprises the properties known as superconductivity and superfluidity, both of which are associated with the behaviour of matter at very low temperatures. A superconductor presents no resistance to the flow of electric current so that if such a current is set up in a loop of superconducting wire it will continue to flow for ever, even though there is no battery or other source of electrical potential to drive it. Similarly, a superfluid is a liquid that can flow along tubes or over surfaces

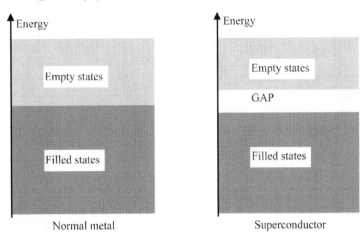

Fig. 7.2 In a normal metal the electrons fill a band of energy states up to a maximum level. Resistance to the flow of electric current arises when an electron is scattered out of a state near the top of the full band into an empty state of similar energy. In a superconductor a gap opens up between the full and empty states so that scattering becomes impossible and current flows without resistance.

without friction. Many metals exhibit superconductivity if cooled to a low enough temperature (typically a few kelvins – i.e. a few degrees above absolute zero). The best-known superfluids are the two isotopes of helium, ^{4}He, which becomes superfluid below about 2 kelvins, and ^{3}He, which exhibits superfluidity only at temperatures less than 0.1 kelvins. The details of the quantum theory of superconductivity and superfluidity are complex, but they both depend on the quantum properties of a macroscopically large number of particles behaving in a collective manner, as we will now try to explain.

To understand superconductivity we must first learn something about electrical conductivity in ordinary metals. Electric current is carried by electrons that are free to move throughout the metal. Resistance arises when the electrons collide with obstacles (such as impurities or defects in the metal structure) and bounce off in a random direction so reducing the current flow. It turns out that these scattering processes are possible only if an electron can be moved from a state with a particular energy to an empty state of similar energy. The energy states in a metal form a continuous band, which can be filled up to a maximum level above which the states are empty (see Figure 7.2). It follows that only those electrons whose energies are near this maximum can be scattered by obstacles. The key to superconductivity is that in some substances the electrons attract each other by a force that is greater than the usual electrostatic repulsion.[3] It can be shown that the effect of this attraction is to produce a gap between the filled and empty states (Figure 7.2). There are now no accessible empty states for the electrons to enter, so they cannot be scattered by obstacles and electric current flows without resistance. The size of the energy gap decreases as the temperature rises, up to a 'critical temperature' above which the metal loses its superconductivity and behaves normally. Until a few years ago, this critical temperature was less than about 30 degrees above absolute zero in all known materials, but in 1986 a new set of materials, known as 'high-temperature superconductors', was discovered with critical temperatures around 100 kelvins.

What is important for our discussion is that in a superconductor the 10^{20} or so electrons all occupy the same quantum state spanning the whole piece of superconductor, which is typically several centimetres in size. The experimental observation of superconducting properties seems therefore to confirm the application of quantum physics to such macroscopic systems and we might think that this is evidence for the absence of spontaneous collapse and therefore close this way out of the quantum measurement problem.

On closer examination, however, it appears that the many-particle loophole is not actually closed just by the existence of supercurrents. Although these confirm the existence of quantum states in which many particles behave coherently, to be relevant to the measurement problem we would have to demonstrate that two such states could be combined coherently in a manner analogous to that suggested for the two pointer oscillations in the previous section. This point was made in the early eighties by the theoretical physicist A. J. Leggett, who suggested that it might be possible to test such a possibility using a particular kind of superconducting device. This proposal has been the subject of considerable attention by a number of theoretical and experimental research groups in this field.

First think of a current flowing clockwise around a ring of superconductor, as in Figure 7.3(a). Because it is a superconductor, such a current will flow forever without resistance unless it is disturbed. Similarly, a current could flow in an anticlockwise direction as in Figure 7.3(b). One important result of a detailed analysis of the quantum physics of such superconducting systems is that these currents can take on only particular 'quantised' values.[4] If we take these states as analogous to those of the swinging pendulum, the question of quantum coherence versus spontaneous collapse could be tested

---

[3] The nature of this attractive force lies in distortions in the arrangement of the charged ions in the solid caused by the presence of the free electrons.

[4] The values of these quantised currents are such that the total resulting magnetic flux in such a ring is always a whole number times $h/(2e)$, h being the quantum constant and e the electronic charge.

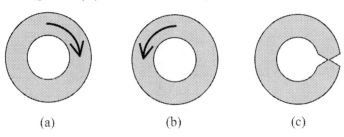

Fig. 7.3 A ring of superconductor can be placed in a state where it carries a current flowing clockwise as in (a) or anticlockwise as in (b). If it could be maintained in a superposition of these two states, then spontaneous collapse would be falsified in this case. A ring with a weak link is shown in (c).

by attempting to put the superconductor into a quantum superposition of such states. If we can do this and if we can observe how the system evolves in time, we can in principle test whether it remains in a superposition state or undergoes spontaneous collapse.

It is known that when electrical currents flow, their motion gives rise to associated magnetic fields; in the case of a current flowing round a ring, the resulting field is similar to that from a magnet of the same size and shape. Also, if an externally generated magnetic field is applied to a superconducting ring, currents are induced to flow in it. These properties can be exploited to induce a particular current configuration in a superconducting ring and to measure its state as a function of time. In addition, to investigate the quantum behaviour of this system we have to modify the superconducting ring by the inclusion of a 'weak link'. This is a point where the ring is made very thin, so that a supercurrent is only just able to flow through it; a superconducting ring with these properties is known as a 'superconducting quantum interference device' or SQUID (see Figure 7.3(c)). It turns out that in a SQUID the clockwise and anticlockwise states can be weakly coupled together. An initial current superposition can then be constructed by applying accurately controlled magnetic fields to the SQUID, and its evolution in time monitored by accurate measurements of the magnetic field associated with the current. Two different kinds of experiment can be carried out to investigate its quantum properties. The first is known as macroscopic quantum tunnelling: this involves setting the ring into a state with (say) a clockwise current, and observing it to see if it makes a transition into an anticlockwise state of lower energy (Figure 7.4(a)). If it does, this can be interpreted as evidence for the ability of the system to 'tunnel' through the energy barrier separating the two states. The second test (Figure 7.4(b)) is to place the system in a

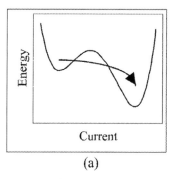
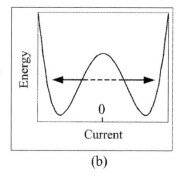

(a)   (b)

Fig. 7.4 (a) In macroscopic quantum tunnelling, a ring undergoes a quantum transition from one potential well to another by 'tunnelling' through the energy barrier that separates them. (b) In macroscopic quantum coherence, the state of the superconducting ring is a superposition of two states that correspond to opposite directions of current flow. Both processes would be disrupted by spontaneous collapse if this occurred.

superposition of two current states and observe how this evolves: this is known as 'macroscopic quantum coherence'.

Although only the second experiment would be a direct test of spontaneous collapse as discussed above, much of the early work in this field consisted of experiments on quantum tunnelling. The reason is that experiments testing macroscopic quantum coherence are technically very much more difficult. However, as the theory underlying quantum tunnelling is closely related to that for macroscopic coherence it might seem perverse if spontaneous collapse were to be limited to the latter context. Nevertheless, the possibility of spontaneous collapse could not be definitively ruled out without a clear demonstration of macroscopic quantum coherence.

In practice, the experiments are difficult to do. In both cases, there are problems in controlling the strength of the weak link to ensure that the energy barrier is the right height: if it is too high then tunnelling through it is virtually impossible while if it is too small the separate current states are not clearly defined. In addition, the experiments have to be very carefully isolated from external disturbances ('noise') that might mimic any effects expected from spontaneous collapse. However, these difficulties have been overcome by the application of considerable experimental ingenuity. Macroscopic quantum tunnelling experiments were first performed in the mid 1980s, while macroscopic quantum coherence was investigated and observed in the late 1990s. All experiments performed so far confirm the predictions of quantum physics and provide no evidence for spontaneous collapse.

Superconductivity has demonstrated quantum coherence for a system containing 10^{20} particles, so does this mean that GRW theory has been falsified? It turns out that this is not quite the case. Although the current-carrying states in a superconductor are macroscopic and can be placed in a superposition, they do not occupy different regions of space. It is a reasonable extension of GRW theory for it to predict that a SQUID should spontaneously collapse into a state of definite current, but it is possible that the model only really applies to an object like a pointer that is in a superposition of states that are separated in space. We conclude that there is now experimental evidence for the quantum superposition of macroscopically distinct states but that this does not completely eliminate the possibility of spontaneous collapse as a solution to the quantum measurement problem.

There is an additional reason why a breakdown of quantum theory in the macroscopic situation may be unlikely. This is that if a macroscopic object were to behave completely classically, in the sense that it was not subject to the Heisenberg uncertainty principle (see Chapter 1), it could have a precise position and momentum simultaneously. It could then be used to make measurements on *microscopic* objects that are more precise than quantum physics allows. This can be understood if we refer to a thought experiment initially discussed by Bohr and Einstein as part of their long debate on the nature of quantum theory (see Chapter 4). They discussed a form of the standard two-slit interference described in Chapter 1 (Figure 1.2) in which the initial slit and screen are mounted on some extremely light springs, as shown in Figure 7.5. The purpose of this first slit is to define the starting point of the light beam, and it is important to note that if it is too wide then the path difference between light waves passing through different parts of it can be greater than a wavelength and as a result the two-slit interference pattern is washed out. Einstein suggested that if the first slit were to deflect a photon towards the upper of the two slits in the second screen then the first screen would recoil downwards, while the opposite would happen if a photon were deflected downwards. It would therefore be possible in principle to tell through which slit the photon would subsequently pass by measuring this recoil. Bohr countered this by pointing out that any such measurement would be tantamount to a measurement of the speed and hence momentum of the recoiling screen. The Heisenberg uncertainty principle would then require there to be a corresponding uncertainty in the screen's position. This would result in an effective broadening of the first slit that would be large enough to destroy the interference pattern. Wave–particle duality is therefore preserved.

Of course this experiment was and is a 'thought' experiment that would be quite impossible in practice, but it is conceivable that the principles involved could be tested by experiments on the delocalisation of flux in SQUIDs. If, however, macroscopic objects like SQUIDs or screens carrying

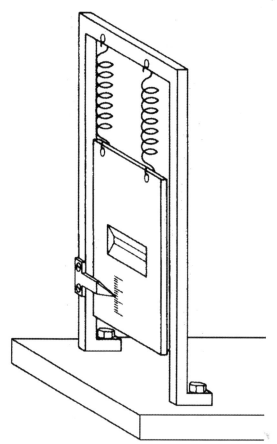

Fig. 7.5 One of the diagrams used by Einstein and Bohr to illustrate their debate about the meaning of quantum physics. If such a
device were used as the initial slit of a two-slit interference experiment
(see Figure 1.2) then the path taken by a photon could in principle
be deduced from the direction of recoil of the slit as indicated by the
pointer. Bohr showed that provided the slit obeys the laws of quantum
physics, such a measurement would destroy the interference pattern.

slits are not subject to quantum physics and if they really have precisely
measurable position and momentum at all times, Bohr's argument no longer
holds. The possibility of the simultaneous measurement of incompatible
properties such as H/V and ±45° photon polarisations (e.g. by observing
the recoil of the first calcite crystal in Figure 4.2) is then opened up, as is
the possibility of similar 'non-interfering' measurements on the photon pairs

in an Aspect experiment. These latter measurements would then have to be subject to Bell's theorem and would therefore give results different from those obtained by conventional means.

As soon as one crack of this kind appeared in the quantum measurement scheme the whole edifice would be in danger of collapse, and a theory that would resolve the measurement problem in the macroscopic situation would also have to avoid breaching the uncertainty principle in this way. GRW theory avoids this problem by postulating that collapse is never to an exact position but to a volume of dimensions much smaller than the separation between the initial delocalised pointer states yet large enough to avoid breaching the uncertainty principle. This of course complicates the theory, which removes some of its appeal. In any case, unless and until experimental evidence emerges to support the principle of spontaneous collapse, most scientists will believe that such a resolution of the measurement problem is very unlikely.

8 · Backwards and forwards

In the last chapter, we considered the possibility that an object such as the pointer of an apparatus designed to measure H/V polarisation (see Figure 7.1) might in principle be forbidden from being in a quantum superposition if it was large enough. A microscopic system, such as a photon polarised at 45° to the horizontal, would then collapse into an h or a v state as a consequence of such a measurement. To test this idea, we considered how we might detect a macroscopic object in such a superposition and we found that this was very difficult in practice. The oscillating pointer of Figure 7.1 is very sensitive to random thermal disturbances and as a result is almost certainly going to swing in one direction or the other, rather than being in a superposition of both. Macroscopic superpositions have indeed been observed in SQUIDs (see the last section in Chapter 7), but only after great care has been taken to eliminate similar thermal effects.

In the context of the last chapter, the randomness associated with thermal motion was considered as a nuisance to be eliminated, but might it instead be just what we are looking for? Perhaps it is not that thermal effects prevent us observing quantum superpositions, but rather that such states are impossible *in principle* when thermal disturbances are present. This could provide another way out of the measurement problem. If we are able to adopt this point of view consistently, we should be able to make the distinction between pure quantum processes and measurements without directly referring to the size or number of particles in a system. Thus an ideal oscillating pointer or a superposition of currents in a SQUID would be in the same category as the 45° photon passing through the H/V apparatus. All these systems could be considered to be in a superposition state until sufficient random thermal motion was involved, after which collapse into a state corresponding to one or other of the possible outcomes would occur. This approach to the measurement problem will be the subject of the present and the following two chapters.

Before proceeding further, we remind ourselves of the difference between an 'in principle' and an 'in practice' solution to the measurement problem. As far as practical consequences are concerned, the assumption that a measurement always corresponds to an irreversible change is a pretty infallible recipe. Thus, as was pointed out in Chapter 7, physicists in practice apply the Copenhagen interpretation, knowing perfectly well when a measurement has been made and what the distinction is between a quantum system and a measuring apparatus. As we shall see shortly, this distinction is effectively the same as that between a situation where the quantum system is isolated from external disturbance and one where thermal effects are of importance and lead to irreversible changes. The first question to be addressed is under what conditions the changes we call 'irreversible' really are that, rather than just processes that could be reversed if we tried hard enough.

However, a second question arises. If we accept the idea that irreversibility is what its name implies, how can this be used to solve the measurement problem? As we discussed in Chapter 4, when the standard rules of quantum physics are applied to the apparatus measuring the H/V polarisation of a 45° photon the predicted state is a superposition of the H and V outcomes. Even if the irreversibility is perfect, so that this superposition can never, even in principle, be observed, the above conclusion is not automatically falsified. We can safely postulate collapse into H or V without fear of subsequent difficulties, but this process still has to be put in 'by hand' and if we don't do so we are essentially back to many worlds. If the theory is to predict collapse, then it will have to include a mechanism for this to happen, so that one or other of the measurement outcomes is 'actualised'. We will return to this question of actualisation in Chapter 10, but for the moment we will concentrate on the first question, which is the nature of irreversibility and the conditions under which it lives up to its name.

Indelible records

Another way of emphasising the importance of irreversibility is to note that an important feature of any real measurement is that an 'indelible' record is made. By this we do not necessarily mean that the result is written down in a notebook or even that the apparatus is connected to a computer with a memory (although either process would constitute the making of a record) but simply that something somewhere in the universe will always be different because the result of the measurement has some particular value. Thus the photon passing through a detector causes a current to pass which may cause a light to flash or a counter to click or a pointer to move to a new position. But even if it doesn't, the current will have caused some slight heating of the air around it, which will have moved a little differently from how it would have done otherwise. Any such change will leave its mark, or record, even if it is so small as to be practically undetectable.

Conversely, a quantum process that involves no interaction with its environment leaves no record of its occurrence at all. When a 45° polarised photon passes through the H/V polariser but then has its original state reconstructed, as in Figure 4.2, there is, almost by definition, no record of which H/V channel it passed through. In the presence of a detector, however, the original

45° state can be reconstructed only if at the same time we 'erase the record' by returning the counter and everything with which it has interacted to their original states. To resolve the measurement problem, perhaps we need only postulate that such a process is impossible and that the 'records' created by measuring instruments cannot be erased but are *indelible*. This may not seem to be a particularly radical postulate – after all we have seen that the probability that all the particles in the detector and its surroundings will return to exactly their original positions is extremely small. However, we would be making a profound distinction between quantum processes, which preserve their full potentiality for reconstructing the original state, and measurement processes, in which a record is made and this potentiality no longer exists.

Irreversibility

In our discussion of the swinging pointer in the last chapter, we saw that it was the random thermal motion of the atoms which in practice prevents the two pendulum motions cancelling out and that similar processes make a superposition of superconducting current states in a SQUID very difficult to observe. This suggests that we should look to the branch of physics dealing with such thermal phenomena for a possible understanding of the distinction between processes in which records are made and those which can be completely reversed, so that no trace of their occurrence remains. The branch of physics that discusses such phenomena is known as *thermodynamics* and it does indeed contain just such a distinction between what are known as *reversible* and *irreversible* processes. This is contained in what is known as 'the second law of thermodynamics', a law which had more publicity than most laws of physics ever achieve when C. P. Snow in his famous essay on 'the two cultures' suggested that some understanding of it should be expected of any non-scientist claiming a reasonable knowledge of scientific culture.

There are a number of different ways of stating the second law, ranging from 'the entropy of the universe always increases' to 'nothing comes for nothing', but for our purposes the basic idea is probably best summarised by the statement that 'any isolated system always tends to a state of greater disorder'. As an example, consider what happens if we have some gas in a container with a partition in the middle (Figure 8.1). Initially we suppose that there is gas only on one side of the partition (achieved by, for example, pumping all the gas out of the other side). If we now open the partition, what will happen? Obviously, because the molecules in the gas are in constant motion, there is a high probability that they will quickly spread out so as to fill the container uniformly and, if we leave the gas alone, we would never again expect to find it all on one side. Because the molecules have more room to move around in the whole container than when confined to one part of it,

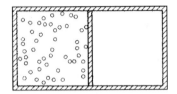 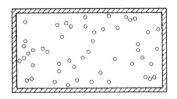

Fig. 8.1 If the barrier confining a gas to the left-hand half of a container is removed, the gas will spread out to fill the whole vessel. The reverse process in which the gas spontaneously moves into one half of the vessel is very improbable and never observed.

they are more disordered in the former case, which is therefore favoured by the second law. Another example is the melting of a piece of ice when placed in warm water: the molecules can move more freely and thereby become more disordered in the liquid than when trapped in the highly ordered structure of the ice crystal. The second law therefore allows this change, while the converse process where a part of the warm water freezes spontaneously is forbidden and, of course, never observed.

Because they apparently happen in one direction only, processes such as those described are known as *irreversible* processes. In contrast, *reversible* processes are those that can happen in either direction. Everyday examples of apparently reversible processes include the swinging of a clock pendulum or the spinning of a wheel. Reversible and irreversible processes can usually be distinguished from each other by imagining a film to be made of the process, which is then run backwards. If the events still appear physically reasonable, they are reversible, and if not, they are irreversible. An educational film produced along these lines shows a bouncing ball (reversible) and a piece of paper being torn (irreversible).

There are two important ideas in this area that at first sight seem contradictory. The first is that all macroscopic processes are actually irreversible, even if they appear reversible, and the second is that all the underlying microscopic processes are actually completely reversible! We shall discuss each of these statements in turn and then try to resolve this apparent paradox; when we do so, we shall find that the resolution of the measurement problem may not be as straightforward as we had hoped.

Why do we say that all macroscopic processes are really irreversible? The crucial idea behind this statement is that it is never possible to isolate a macroscopic body completely from its surroundings. Thus, although the swinging pendulum appears to be behaving reversibly, careful examination will show that its motion is actually slowing down, albeit gradually, due to drag forces from the surrounding air and friction at the pivot bearings.

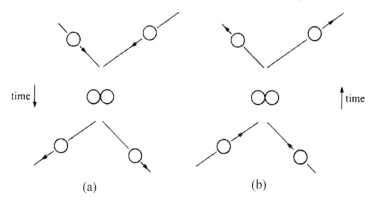

Fig. 8.2 Elementary dynamical processes, such as the collision of two spheres, appear just as physically reasonable if their time order is reversed.

Alternatively if, like a clock pendulum, its amplitude remains constant, this implies that it is driven by some kind of motor; examination of this power source will reveal irreversible processes such as an unwinding spring or a discharging battery. One apparently very reversible large-scale process is a moon orbiting a planet or a planet orbiting the sun. These objects are passing through the vacuum of space where there is practically nothing to exert drag forces and their motion has been going on for millions of years without stopping. Nevertheless it is known that there are frictional forces even in this situation: for example, the tides on the earth caused by the moon and the sun dissipate energy, which comes from the orbital motions so that these are slowed down. As a result the average distance between the moon and the earth increases by about 38 millimetres each year, (measured by laser-ranging experiments) and the earth's rotation slows by about 0.8 seconds per year (confirmed using atomic clocks). This part of the moon's motion is therefore irreversible in the same way as the other examples discussed: for the moon to move towards the earth instead of away from it, the tides would have to supply energy to the moon, rather than the other way round, and this would breach the second law.

We turn now to the second statement in the paradox: that all fundamental microscopic interactions are reversible. This arises because the laws of mechanics and electromagnetism do not depend on the direction (or 'arrow' as it is sometimes called) of time. A simple situation illustrating this is a collision between two molecules, as shown in Figure 8.2(a) and its time reverse, Figure 8.2(b). We see that both these processes are equally acceptable physically and that there is nothing in the motion to tell us in which

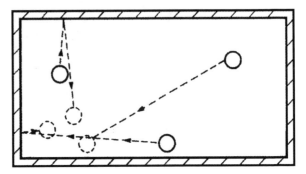

Fig. 8.3 Although a gas is never seen to spontaneously concentrate itself into one half of its container, this is not forbidden by the laws governing atomic collisions. The diagram shows three particles that are initially widely separated (solid circles) moving in such a way they move together into the bottom left-hand corner (broken circles).

direction time was actually flowing. This is true of all physical events occurring at the microscopic level.[1] How then does it turn out that the behaviour of large-scale objects is always irreversible when the motion of their atomic constituents follows these reversible macroscopic laws? The standard answer to the problem of how macroscopic irreversibility arises from microscopic reversibility is that the irreversibility is an approximation, or even an illusion. To understand this, consider again the case of the container of gas (Figure 8.1) whose partition is removed so that the gas then fills the whole vessel. If we were to concentrate on a single gas molecule we would find it undergoing a series of collisions with the walls of the vessel and with other molecules, each of which is governed by reversible laws. Suppose that at a certain time we were able to reverse the direction of motion of every molecule in the gas. Because the mechanical laws governing the collision work just as well 'in reverse', each molecule would retrace the path it had followed and the molecules would all be back in the left-hand part of the container after a further time, identical to that between the opening of the partition and the reversal. We could close the partition and so restore the original state of the system. An example of this process being followed by a 'gas' of three molecules is illustrated in Figure 8.3. They are initially spread around the container, but they happen to move in such a way that they come together in one part of it after a short time. Of course stopping and reversing the molecular motion is not possible in practice, becoming rapidly more difficult as the number of particles in the system increases. However, it is apparently always

---

[1] Apart from one or two processes involving particular subatomic particles, which do not need to concern us.

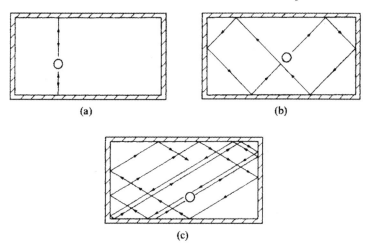

Fig. 8.4 A single particle moving in a rectangular container can undergo different kinds of motion depending on its starting conditions. In (a) and (b) the motion repeats cyclically after a few collisions. Far more likely is a path such as that shown in (c) which never returns to its precise starting point. In this latter case the ergodic principle implies that the particle will eventually occupy all possible dynamical states.

possible in principle. Moreover, a more detailed analysis of the motion leads us to believe that, even in the absence of intervention, a physical system will always revisit its initial state if we wait long enough.

To understand this last point, we first consider the very simple example of a single molecule moving inside a flat rectangular area and bouncing off its boundaries, as in Figure 8.4. If it is set off moving parallel to one of the sides of the rectangle then it will bounce up and down along a line indefinitely. Every other time it passes the starting point, it has the same speed in the same direction as it had originally, so in this case the initial state recurs very quickly. Similarly, if the molecule starts off from a point on one side at an angle of exactly 45°, it will trace out the path shown in Figure 8.4(b) and continue to follow this course indefinitely. Other similar repeating paths can easily be devised. However, in the case of a more general starting angle the path will be very complex and an imaginary line drawn by the particle will eventually fill the whole rectangle.[2] The starting conditions leading to motion of this kind are infinitely more probable than those leading to simple

---

[2] It can be shown in a reasonably straightforward way that the simple repeat occurs only if the tangent of the starting angle is a rational number – that is if it is equal to n/m where n and m are whole numbers.

repeats so, in general, we can assume that the particle will have occupied every possible position on the rectangle, including that corresponding to its initial state, if we wait long enough.

We can generalise this result to describe the behaviour of a more complex system, such as the gas discussed earlier. We now have a large number of particles moving and colliding with each other and the walls of the container. The same arguments that we applied to the single-particle case lead to the conclusion that that such a system, if left to itself, will sooner or later pass through every possible configuration consistent with the principle of conservation of energy. This statement is known as the *ergodic principle* and is generally assumed to be true for any physical system such as a gas of molecules. Of course, such a description of the physical state of the system may require the knowledge of the position and velocity of each of the 10^{23} or so molecules in the gas, so the number of possible states to be passed through is immense. If the ergodic principle is correct, however, all possible states will eventually occur and some of these states, even if they are not precisely identical to the initial state of the gas, will be effectively indistinguishable from it. It follows therefore that the gas will revisit a state that is arbitrarily close to its initial state if we wait for a long enough time. This fact was first realised in the nineteenth century by the French physicist Henri Poincaré: the return of a thermodynamic system to an earlier state is known as a 'Poincaré recurrence' and the time it takes for this to happen is known as the 'Poincaré cycle time'.

It is arguments such as these that lead to the belief that irreversibility is an illusion. Although the gas expands to fill the empty box there is always some chance, however small, that it will return to its initial state with all the molecules in the left-hand half of the container, and this will certainly happen if we wait long enough. Of course we must expect to wait a very long time as estimates of Poincaré cycle times for typical macroscopic systems come out at many millions of times the age of our universe. However, the Poincaré time is just an average: there is an extremely tiny, but finite, chance that one day all the gas molecules in a box or a room may move into one part of it, with very uncomfortable consequences for any of us who happen to be in the room at the time!

Irreversibility and measurement

It should now be clear why an appeal to thermodynamic irreversibility does not immediately resolve the measurement problem. It is certainly true that pure quantum processes are strictly reversible in a thermodynamic sense, while measurements always involve thermodynamic irreversibility. However, if the ergodic principle and the idea of Poincaré recurrence are correct, this irreversibility is an illusion arising from the fact that we cannot observe any large-scale process for a long enough time to expect to see a Poincaré recurrence. If we could make such an observation, the atoms in the swinging pointer in Figure 7.1 *would* eventually happen to be in exactly opposite positions when swinging in the two directions. As a result the two motions would cancel out, and the polarisation state of the photon *would* still be 45°, and neither h nor v. Even if some other record had been made of which way the pointer was swinging, this must also be constructed out of materials that are subject to the laws of thermodynamics and so the whole lot together can be considered as a single thermodynamic system and subject to an eventual Poincaré recurrence. If there is no point at which we can say that the measurement has finally been made, then the Copenhagen interpretation insists that we must not ascribe any reality to the unmeasured quantities (the photon polarisation, the direction of the pointer swing or indeed anything else!) and we are still impaled on the horns of the measurement problem.

To make progress in this direction we are going to have to find some reason why the ergodic principle and the ideas of Poincaré are not applicable to a quantum measurement. One suggestion that has been made is to note that no real thermodynamic system is ever completely isolated from its surroundings. Thus the atoms in a gas interact with external bodies, if only through the weak gravitational interactions that operate between all objects. Although such influences are extremely small, it has been shown that very tiny perturbations can have a large effect on the detailed motion of a complex many-particle system such as a gas or, indeed, a pointer. Thus the gravitational effects of a small change in a distant galaxy could affect the behaviour of a container of gas on the earth to such an extent as to make a major change in the time of a Poincaré recurrence. Probably no complex system can be considered as having returned to a state it was in earlier unless its whole environment is also the same as it was previously – because, if this is not the case, its future evolution will not be the same as it was the first time round. We could therefore suggest that a measurement represents an irreversible change in the whole universe and simply postulate that the laws of quantum physics are applicable to any part of the universe, but not to the universe as a whole. We might even argue that it is meaningless to consider the possibility of the universe returning to a state it had occupied previously as this would imply that the state of all observers in the universe would have to be similarly restored and there is no way we could know that anything had occurred or that time had passed. There are two objections to this. First, incredible as it may seem, it does appear as if quantum theory can be applied to the universe as a whole. Theoretical studies of its development during the very early stages of the 'big bang' rely on just this postulate and produce results that appear to explain some of the observed properties of the universe today. Secondly, it is difficult to decide when an irreversible change in the whole universe has occurred. All physical influences, including gravity, travel at the speed of light or slower so it would take many years before the change associated with a measurement was registered in the distant parts of the universe. We would have problems if we could refuse to ascribe reality to an event for the first few thousand million years after its occurrence!

There is, however, a completely different approach to the problem and this is simply to postulate that the ergodic principle is not correct and that a Poincaré recurrence of the type required to reverse quantum measurements just does not happen. After all, no breach of the second law of thermodynamics has ever been observed in practice and the whole argument rests on some really quite sweeping hypotheses. Could we not find some aspect of the nature of the thermodynamic changes that occur in measuring processes which would clearly distinguish them from the kind of process for which reversibility can be envisaged and to which pure quantum theory can be applied? This possibility will be discussed in the next two chapters, where we shall see that it suggests a solution to the measurement problem that could be thought of as more down to earth than any we have discussed so far. However, we shall find that it also implies a revolutionary change from conventional forms of description for the physical universe.

9 · Only one way forward?

In this chapter we ask if there is some aspect of the nature of the thermodynamic changes that occur in measuring processes that could clearly distinguish them from the kind of process for which reversibility is a possibility and to which pure quantum theory can be applied. This idea has been suggested on a number of occasions, and was developed in the 1980s by Ilya Prigogine, who won a Nobel prize in 1977 for his theoretical work in the field of irreversible chemical thermodynamics. The starting point of the approach by Prigogine and his co-workers is a re-examination of the validity of the ergodic principle, which leads to the idea of the Poincaré recurrence. We made the point in the last chapter that, in the simple case of a single particle confined to a rectangle, unless the starting angle has a special value, the particle trajectory will fill the whole rectangle and the particle will sooner or later revisit a state that is arbitrarily close to its initial state. By this we meant that, although the initial and final states cannot in general be precisely identical, we can make the difference between the initial and final positions and velocities as small as we like by waiting long enough. The implicit assumption is that the future behaviour of the system will not be significantly affected by this arbitrarily small change in state; its behaviour after the recurrence will then be practically the same as its behaviour was in the first place. Hence, if the particle is part of a quantum system, the recurrence will reconstruct the original quantum state so precisely that we cannot say that any change has really taken place. This idea is probably applicable to a simple system such as a single particle bouncing around a rectangle, but it turns out that it has only limited validity in the case of more complex systems.

As an example of a physical system whose future behaviour can change drastically as a result of an arbitrarily small change in its initial state, we consider a simple pendulum constructed as a weight at the end of a rod mounted on a bearing, so that it can rotate all the way round (Figure 9.1). This pendulum has two possible, quite distinct, types of motion. If it is pulled a bit to one side and released, it will swing backwards and forwards like a clock pendulum, but if it is given a large initial velocity it will swing right over the top and continue rotating until it slows down. The important point is that there is a critical speed below which the pendulum will swing up to the top and back again and above which it will go over the top and rotate. It follows

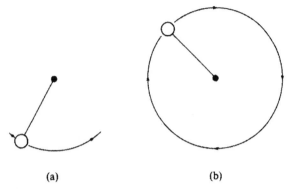

Fig. 9.1 A simple pendulum can undergo two distinct types of motion,
either oscillation as in (a) or full rotation as in (b).

that the future behaviour of two such pendulums, whose only difference
is that their velocities are higher and lower than this critical velocity by
arbitrarily small amounts, will be completely different. Consider now what
would happen if such a pendulum were part of a thermodynamic system. For
example, we might imagine a very small version of it in a container along
with a number of gas molecules, so that from time to time the molecules
strike the pendulum giving it various amounts of energy. If at some point the
pendulum has very nearly enough energy to start rotating, it is possible that
the system could revisit this state, the gas molecules being in practically the
same places and moving with almost the same speeds but the energy given to
the pendulum being just slightly greater than before. The pendulum would
now start rotating, and its whole future behaviour and that of the molecules
colliding with it would be quite different from that which followed the initial
state. It should now be clear that there are difficulties in applying the ergodic
principle and the idea of a Poincaré recurrence to a system like this: however
similar the initial and subsequent states of such a system may be, their future
behaviour may be completely different.

It might seem that a situation like the above is so unusual as to be
practically irrelevant. However, the details of the motion in many- particle
systems are extremely complex and such instabilities are actually quite com-
monplace. For many systems, *any* arbitrarily small change in the starting
conditions drastically alters its future behaviour. The effect of such instabilities on a physical system is sometimes known as 'strong mixing' and systems
that are liable to this behaviour are often called 'chaotic'. It has been shown
that a 'gas' composed of as few as three hard-sphere particles behaves in this
way, so it is very likely that strong mixing is a feature of most, if not all, real
thermodynamic systems.

Following Prigogine, we first consider the relevance of strong mixing to our understanding of classical physics and return to the quantum case and the measurement problem a little later. The first point to note is that, even although the detailed future behaviour of the component particles of a chaotic system is unpredictable, this does not mean that all its physical properties are unmeasurable. In particular, traditional thermodynamic quantities such as the temperature and pressure (if it is a gas) are perfectly well defined. Moreover, the second law of thermodynamics is strictly obeyed, the system tending to a state of greater disorder even more rapidly than it would do if subject to the ergodic principle. It is the underlying dynamical parameters, such as the positions and velocities of the component molecules, that change chaotically whereas the thermodynamic quantities (which are traditionally thought to be derived from the microscopic substructure) are well behaved. These facts led Prigogine to suggest that our traditional way of thinking about thermodynamic systems is wrong: if strong mixing is present, we should consider the thermodynamic quantities to be the primary reality and the allegedly more fundamental description in terms of microscopic structure to be secondary. He suggested that it would then be possible to treat the second law of thermodynamics in the way its name suggests: as a *law* that is always followed rather than a statistical rule that would always be eventually broken by a Poincaré recurrence.

The consequences of this way of thinking are profound even before we consider its application to quantum theory. If we follow Prigogine's approach, indeterminism becomes an implicit part of *classical* physics: our inability to predict the future motion of the components of a many-particle system is no longer to be thought of as a limitation on our experimental or computational ability, but as an inevitable consequence of the laws of nature. Instead of taking as fundamental the laws that refer to microscopic reversible processes (with macroscopic irreversible behaviour as an approximation or illusion) it is the irreversible laws that should be taken as fundamental and reversibility that is the approximation. Prigogine postulates a kind of uncertainty principle linking the two types of description: if the thermodynamic description is appropriate, then precise measurement of the dynamical variables is impossible, while a simple system that can be described dynamically (like the single particle on the rectangle) possesses no definite thermodynamic parameters. We might even go so far as to extend the ideas of the Copenhagen interpretation to this field, and suggest that it is just as meaningless to ascribe reality to the positions and velocities of the individual particles in a system subject to strong mixing as it is to talk about the temperature of a single isolated particle. It is interesting to observe some of the ideas of quantum physics showing up in this way of thinking about a completely classical system.

The idea that it is reversibility that is the approximation to irreversibility rather than the other way round may be more consistent with our view of the universe as we see it. Thus (still ignoring quantum ideas, to which we shall be returning very soon) all real systems are visibly subject to the laws of thermodynamics and are 'running down'. Anyone who claimed to observe a Poincaré recurrence in a thermodynamic system would be disbelieved as much as any other reporter of an alleged miracle. If the second law is fundamental then this running down is predicted and the Poincaré recurrence is not possible. The time to run down varies very much from system to system. Thus the block of ice in warm water melts in a few minutes, but it will be many millions of years before the solar system is consumed by the sun, even longer before the galaxy collapses and longer still before the universe reaches some final equilibrium state. Our ability to predict the future behaviour of physical systems seems to be greater for the large-scale parameters of large systems. Thus we are unable to predict the future trajectory of a molecule in a gas more than a few collisions ahead, but we can work out to considerable precision how the planets will be moving round the sun in some thousands or even millions of years' time. Eventually, however, the behaviour of the solar system will become unpredictable and chaotic. In the very long term, the only things we can predict are the global parameters of the universe (its overall size, mass, temperature, etc.). The fact that these cannot be accurately estimated at the present time is probably due to the inaccuracy of our data rather than fundamental limitations of the type we have been considering.

Return to the measurement problem

Everything we have said so far in our discussion of strong mixing has referred to classical systems, but it is now time to return to the main point at issue, the quantum measurement problem. Superficially at least the solution is very simple. If we postulate that the second law is fundamental for all physical systems that are subject to strong mixing, so that a Poincaré recurrence can never occur in such cases, then it is a short step to say that all quantum measurements are in practice carried out with apparatus of this kind, and the measurement chain is broken whenever strong mixing becomes involved. This is indeed Prigogine's solution to the measurement problem, but he also goes further than this and explores some of the consequences this picture has for our view of the quantum world.

In the same way as it was always believed that the fundamental parameters of a classical system are the positions and velocities of its components and that the thermodynamic description is a statistical approximation, the traditional approach to quantum physics is to give primacy to the quantum states of the system and to look on the measurement interactions as secondary. Thus, as we have seen, the whole measurement problem arises when we treat the measurement apparatus in the same way as a quantum object, such as a photon. Prigogine's approach is to reverse this way of looking at things: the behaviour associated with a quantum measurement is taken as the fundamental reality, while pure quantum behaviour is an approximation appropriate only to the special situations where the effects of strong mixing (always present in a measurement) can be neglected.

This new approach also changes the way we look at the fundamental indeterminacy traditionally associated with quantum physics. To see this, we note first that the behaviour of a pure quantum system in the absence of a measurement is actually quite predictable: the 45° photon passing through the H/V polariser does not pass through one channel or the other but, in some way we find very difficult to model, it passes through both. If it didn't, we wouldn't be able to reconstruct the original state in the way we have so often discussed. Moreover, even in the absence of such a reconstruction, quantum theory describes the photon as being in a well-defined state that is a superposition of an h photon in one channel and a v photon in the other. The indeterminacy arises when we allow it to interact with a measuring apparatus, which results (*pace* many-world supporters) in the random collapse of the state into one polarisation or the other. But we have seen earlier that there is a fundamental indeterminism, quite independent of quantum effects, associated with the strong mixing always implicit in a measurement. Is there then any need to introduce an additional uncertainty associated with the quantum system? Prigogine's answer is 'no'. He has shown that the extension of his ideas to the quantum regime actually introduces extra correlations that reduce the indeterminism associated with strong mixing in the classical situation to a point consistent with the quantum uncertainty principle.

The distinction between irreversible measurements and reversible pure quantum processes makes it a little easier to accept some of the peculiar features of the latter. If there is no irreversible change associated with a quantum event, it is perhaps hardly surprising that we find it difficult to describe such processes in conventional terms. After all, if we knew through which H/V channel the photon had passed then an irreversible change would have occurred somewhere, if only in our brains, and the process would no longer be completely reversible. A pure quantum process occurs only in a parameter, or set of parameters, that are detached from the rest of the universe and leave no trace of their behaviour until a measurement interaction takes place. We should perhaps be more surprised by the fact that quantum theory allows us to say anything at all about the behaviour of a quantum system between measurements than by our inability to make a precise description in this realm.

We see now why these ideas, although apparently quite simple, involve a quite revolutionary change in our thinking about the physical universe. For a long time now the emphasis in physics has been on measuring and understanding the behaviour of the elementary, subatomic particles that are believed to be the fundamental building blocks of nature. It is implicitly, or sometimes explicitly, stated that the behaviour of macroscopic bodies, or even the universe at large, can be understood purely in terms of these elementary particles and the interactions between them. Thus the behaviour of a gas composed of such particles, all moving subject to reversible laws, must also be reversible, and the apparently irreversible changes must be an approximation or illusion resulting from our observation over too short a time scale.

Prigogine completely reverses this way of looking at things. He suggests that it is the irreversible changes that are the really fundamental entities in the universe, and that the idea of microscopic particles moving subject to reversible laws is an approximation, valid only in the very special circumstances where a particle is effectively decoupled from the rest of the universe. Notice that the fundamental concepts are the events or changes rather than the objects that are doing the changing. Prigogine's own words to describe this change of emphasis from 'being' to 'becoming' are as follows.

The classical order was: particles first, the second law[1] later – being before becoming! It is possible that this is no longer so when we come to the level of elementary particles and that here we must *first* introduce the second law before being able to define the entities. Does this mean becoming before being? Certainly this would be a radical departure from the classical way of thought. But, after all, an elementary particle, contrary to its name, is not an object that is 'given'; we must construct it, and in this construction it is not unlikely that *becoming*, the participation of the particles in the evolution of the physical world, may play an essential role.

Although this change of emphasis may seem revolutionary, it is as we have seen, quite consistent with our experience of the physical universe. Any experience we have is certainly of irreversible processes involving strong mixing, if only because the changes taking place in our brain are of this nature. By definition we have no experience of reversible, pure quantum 'events' that are not detected. But this is not to say that Prigogine's approach is in any way subjective in the sense discussed in Chapter 5. What is important is not the fact that pure quantum 'events' have no effect on *us* but the fact that that they result in no permanent change to any part of the universe. The laws of classical physics were set up on the unquestioned assumption that, although events may be reversible, it is always possible to talk about what has happened. However, even Einstein's theory of relativity refers extensively to the sending of signals, which are clearly irreversible measurement-type processes. Perhaps it should not be surprising that, when we try to construct a

---

[1] i.e. the second law of thermodynamics.

scenario that goes beyond the realm of possible observation into the reversible regime, our models lead to unfamiliar concepts such as wave–particle duality and the spatial delocalisation observed in EPR experiments.

Another feature of this way of looking at the measurement situation is that it actually brings us back to something very like the Copenhagen interpretation. Those of us trained in this point of view have just about learned not to attribute 'reality' to unobservable quantities. Thus questions such as whether an object is really a wave or a particle or whether a 45° polarised photon really went through the H or the V channel can in principle never be answered, so there is no point in asking them; the emphasis should always be on the prediction and understanding of the results of measurements made by observers. As we have seen in the last few chapters, the measurement problem arises when we attempt to make a consistent distinction between the observer and the observed object. Prigogine's approach is to distinguish not between these entities but between the nature of the processes. If we identify the idea of a measurement by an observer in the Copenhagen interpretation with an irreversible change in the universe brought about by the onset of strong mixing, we may well have obtained a consistent solution to the measurement problem.

It would be wrong to say that all Prigogine's ideas have become part of the scientific consensus, accepted as orthodox by the general body of physicists; indeed, much of the work done in the field since they were developed in the early 1980s seems to have gone ahead in parallel with his work and makes little acknowledgement of it. Nevertheless, much of his thinking does pass the test of time. Classical chaos has been the subject of considerable study and the fact that chaotic systems are more strongly disordered than would be expected from the ergodic principle is now well established. Even so, the idea of applying the Copenhagen interpretation and the uncertainty principle to classical systems has not caught on. One reason for this may be that there appears to be no fundamental constant to fill the role played by Planck's constant in quantum physics.

In the quantum case, and particularly in the context of measure- ment, the importance of irreversibility is widely recognised. Again, much of this development seems to have been parallel to, rather than following from, Prigogine's work. However, measurement theories that treat irreversibility as fundamental have emerged strongly in the latter years of the twentieth century; one example of this trend is that the branching points in many-worlds theories are often identified with irreversible thermodynamic changes. The next chapter discusses some of these modern developments in more detail.

10 · Can we be consistent?

Irreversibility, strengthened by the idea of strong mixing, has been discussed in the last two chapters. We reached the conclusion that, once such processes have been involved in a quantum measurement, it is in principle impossible to perform an interference experiment that would demonstrate the continued existence of a superposition. It is then 'safe' to assume that the system has 'really' collapsed into a state corresponding to one of the possible measurement outcomes. Does this mean that the measurement problem has been solved? Clearly it has for all practical purposes, as has been pointed out several times in earlier chapters. But it may still not be sufficient to provide a completely satisfactory solution in principle: in particular, we note that we have still not properly addressed the question of actualisation outlined at the end of the first section of Chapter 8.

In the present chapter we discuss an interpretation of quantum physics that was developed during the last 15 or so years of the twentieth century and is based on the idea of describing quantum processes in terms of 'consistent histories'. As we shall see, the resulting theory has much in common with the Copenhagen interpretation discussed in Chapter 4 and when applied to measurement it connects with the viewpoint, discussed in the last chapter, in which irreversible processes are to be taken as the primary reality.[1]

We begin with a short discussion of the meaning and purpose of a scientific theory. A useful analogy is a map that might be used to navigate around a strange city. Though the map is normally much smaller than the physical area it represents, if it is a good map it is a faithful representation of the terrain it is modelling: streets and buildings are related to each other in the same way as they are in reality. Clearly, however, the map is not the same as the terrain it models and indeed is different from it in important respects; for example, it is usually of a different size from the area it represents and is typically composed of paper and ink rather than earth and stones.

A scientific theory also attempts to model reality. Leaving quantum physics aside for the moment, a classical theory attempts to construct a model

---

[1] As we said at the end of Chapter 9, much of this work seems to have been done independently of Prigogine's work. However, the fundamental idea that the irreversible outcome of a measurement should be a central part, or even the starting point, of quantum physics is an important part of the consistent-histories approach.

118

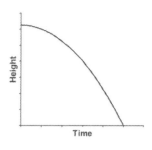

Fig. 10.1 We can use Newton's laws to construct a graph or 'map' of the motion of an apple falling from a tree.

or 'map' of physical events. Consider the simple case of an object, such as an apple, falling under gravity. The apple is at rest, is released, accelerates and stops when it reaches the ground. Such a sequence of events is often referred to as a 'history', and the physicist's map of this history is constructed using Newton's mechanics. Two types of information are required in order to construct this map: the first is our knowledge that all falling bodies accelerate under gravity, while the second is the particular body's initial position and speed. Given these we can calculate how fast the particle will be travelling and how far it will have fallen after a given time, as shown in Figure 10.1. To construct this map, we have done a little mathematics, but, although the real apple is incapable of performing the simplest mathematical calculation, it still falls to the floor in the predicted time.

Another example is the motion of the planet Mercury around the sun, where a construction of a map of the motion requires the application of the equations of general relativity. These were unknown until Einstein's work early in the twentieth century, and they still challenge the understanding of all but specialists in the field. Nevertheless, Mercury has had no problem in following this orbit before the time of Einstein or since! The aim of science is to construct the most detailed and faithful map of physical reality as is possible. This can require extensive use of mathematics, which is used to construct a map of reality; but the map is not reality itself.

Clearly it is essential that we choose a map appropriate to the physical situation that we are addressing. A map based on Maxwell's theory of electromagnetism will be of little use to us if we are trying to understand the fall of the apple under gravity. Even if we do choose a map based on Newton's laws it still has to be appropriate: for example, it must include a representation of the effects of air resistance unless our apple is falling in a vacuum.

When we construct maps of the quantum world, we find that choosing the appropriate map to suit the particular context is all important.

So what kind of map can we construct in the quantum case? The central feature of the consistent-histories approach is that there is no single map of the quantum world. Rather, quantum theory provides a number of maps, and which one we should use in a particular situation, depends on the context, or even on the experimental outcome, and may change as the system evolves in time. Pursuing our analogy a little further, we might say that quantum physics enables us to construct a 'map book' and that we must use the map that is appropriate to the particular situation we are considering. Ignoring the possibility of hidden variables, one thing we know is that a quantum map can never model simultaneously the exact position and speed of a particle or tell us both the H/V and ±45° polarisations of a photon. Maps that purport to do this are termed 'inconsistent', for reasons that should become clearer shortly, and are forbidden in this approach.

As an example, consider again the example of photon polarisation. We know that we can pass a photon through an H/V apparatus and find that it is either in an h state or a v state, or we can pass it through a ±45° polariser and detector and find that it is in either a +45° or a −45° state. Or, indeed, we could orient the polariser in some other direction and there would again be two possible outcomes. In the absence of any such measurements, can we say anything at all about the photon's polarisation? The consistent-histories approach to quantum physics answers with a qualified 'yes' to this question. The isolated photon can be described as being either h or v; as either +45° or −45°, or as being polarised either parallel or perpendicular to some other direction. Each possible state of the photon is denoted as a 'history' and a set of alternatives, such as h or v is known as 'consistent family of histories'. In contrast, if we were to say that a photon is 'either h or +45°', this would constitute an *inconsistent* family of histories because if it has one of these properties, the other is meaningless (see Chapter 4). It is a rule of this interpretation that only consistent histories are allowed. Pursuing our previous analogy, every page in the map-book contains a set of consistent histories; we can choose to use whichever map is most useful to describe the quantum situation, but we must use only one such map at any one time.

In the above example, identifying which histories are consistent is reasonably obvious, but this is not always the case. However, a general test for consistency can be made on the basis of whether the calculated probabilities of the different histories have the essential properties of probabilities. In particular, the joint probability of two histories in a consistent family should be the sum of their individual probabilities. Consider for example, a photon that has emerged from a polariser in a +45° state. If we were to re-measure the ±45° polarisation, we would certainly get +45° again, so the probability of

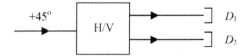

| 1 | +45°, D_1,D_2 | h, D_1,D_2 | $h, D_1{*},D_2$ |
| 2 | +45°, D_1,D_2 | v, D_1,D_2 | $v, D_1,D_2{*}$ |

Fig. 10.2 A photon with 45° polarisation passes through an H/V polariser and is then detected in D_1 or D_2. The table gives the state of the photon and the detectors at each stage of the two possible consistent histories. The asterisk indicates that the detector has registered a photon.

this is unity and the probability of the alternative history ($-45°$) is zero. The joint probability (i.e. the probability of either $+45°$ or $-45°$) is $1 + 0 = 1$. If instead, we were to measure H/V, we would obtain h or v, each with a probability $1/2$ and, as these are the only possible outcomes, the joint probability of h or v is $1/2 + 1/2 = 1$, so h and v also form a consistent family. In contrast, if we apply the standard laws of probability to the outcomes $+45°$ or h, we would deduce that the joint 'probability' of getting either result would be $1 + 1/2 = 1\frac{1}{2}$, which is greater than unity and so cannot be a probability. Such a family is termed inconsistent and is forbidden in the consistent-histories interpretation. We note that the particular cases discussed earlier are therefore consistent with the joint-probabilities rule. The theoretical formalism of quantum physics allows us to calculate probabilities and test for consistency in more general cases, where the number of histories in a family can be much greater than the two considered so far, and also to calculate how the various histories change with time.

To understand how the consistent-histories approach deals with the question of quantum measurement, consider once again a photon polarised at 45° to the horizontal entering an H/V polariser, after which it is detected in one or other output channel (Figure 10.2). Initially, the appropriate map describes a $+45°$ polarised photon moving from left to right, and in this situation there is only one possible history. The photon now moves through the polariser. When it emerges, but before it is detected, we have a choice of consistent families: one would represent the photon in a superposition state while the other would have two histories, in which the photon is respectively either horizontally or vertically polarised. The presence of the detectors means that only the H/V map is relevant to the case where the photon will be detected in one or other channel. (Remember that the interpretation allows us to use whichever consistent family is relevant to the situation we are considering.)

Finally, when the photon is detected in, say, the H channel, the relevant history is that corresponding to this outcome. Beforehand, we do not know which history will describe reality at this final stage, although we can predict the relative probabilities in the standard way (50:50 in the present example; see Chapter 2 for the more general case). This is how indeterminism is built into a model based on histories.

It is possible to combine the histories corresponding to the different stages into histories spanning the whole process and to include the state of the detectors at each stage. In the present example, this leads to two possible consistent histories, as described in the table in Figure 10.2. Once we know the outcome of the measurement, we can retain the consistent history representing it and discard the other. Does this mean that we are basing everything on our conscious experience, as we discussed and criticised in Chapter 5? Not at all! Our knowledge tells us which family we should retain (i.e. which 'map' we should use) but physical reality, which is *not* the map is unaffected.

An important, and sometimes controversial, feature of the consistent-histories interpretation is that when an irreversible change occurs, the separate outcomes (or 'branches' in the language of Chapter 3) form a set of consistent histories. This is despite the fact that they fulfil the joint-probability test mentioned above only if the changes are *completely* irreversible, in the sense discussed in Chapters 8 and 9. We saw in Chapter 8 that any ergodic system always has an extremely small probability of a Poincaré recurrence, and this is reflected in very tiny deviations from consistency in the consistent-histories approach. We must either ignore these or assume that processes such as strong mixing (see Chapter 9) actually reduce them precisely to zero. We can then use a map where the final stage of the above example is one containing histories corresponding to h and v photons emerging through the H and V channels respectively. Using this map we can then interpret the experiment as one with definite possible outcomes to which we can assign probabilities. We note that this approach to the quantum measurement problem gives primacy to irreversible changes such as those associated with measurement. We notice the connection between this aspect of the philosophy underlying the consistent-histories interpretation and Prigogine's ideas discussed in Chapter 9. However, even in the absence of measurement, the consistent-histories approach restricts the possible descriptions of reality to those contained in the relevant consistent families. To this extent it provides at least a partial ontology for the quantum world.

It should also be clear that the consistent-histories approach has much in common with the Copenhagen interpretation discussed in Chapter 4. The fundamental idea underlying the present approach is that our theoretical structure (or map) connects with physical reality *only* when it contains a

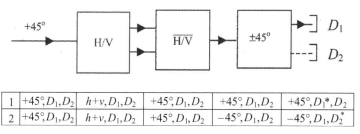

| 1 | $+45°,D_1,D_2$ | $h+v,D_1,D_2$ | $+45°,D_1,D_2$ | $+45°,D_1,D_2$ | $+45°,D_1^*,D_2$ |
| 2 | $+45°,D_1,D_2$ | $h+v,D_1,D_2$ | $+45°,D_1,D_2$ | $-45°,D_1,D_2$ | $-45°,D_1,D_2^*$ |

Fig. 10.3 The action of the first H/V polariser is reversed by the second, so that the initial $+45°$ polarisation is unaltered. This is reflected in the consistent histories: in this example, the second history has zero probability. By '$h + v$' we mean a photon in a state that is a superposition of h in the H channel and v in the V channel.

set of consistent family of histories to which probabilities can be assigned. This includes the Copenhagen principle that reality consists of the results of measurements, but has the advantage of being more general as well as more objective. Niels Bohr's tenet that the experimental arrangement determines the possible attributes that a quantum system may possess is re-expressed in the rule that all histories relevant to a particular situation must form a consistent family. As we have seen, this can be deduced directly from the equations of quantum physics rather than requiring a subjective judgement. Nevertheless, the map is entirely our construct; much of it corresponds to how we have decided to think about the problem, rather than with what may 'actually' be happening. Very much in the spirit of Bohr, we do not attempt to describe processes that are in principle unobservable. However, an important distinction between the consistent-histories approach and the Copenhagen interpretation is that the former does not rely on a definition of measurement. As we have seen, even isolated quantum systems, in which no measurement occurs, can be described in terms of families of consistent histories.

To illustrate these ideas further, we consider two more examples. The first (see Figures 10.3) is the situation where a 45° photon passes through two H/V analysers, oppositely configured so that the splitting caused by the first is undone by the second (as discussed in Chapter 4) and is then subject to a ±45° measurement. The consistent histories in this case are also shown in Figure 10.3. Putting the first history into words, the photon starts in a +45° state, evolves into an $h + v$ superposition and back into a +45° state, neither detector having recorded a photon until the final stage when one detector registers the outcome of the measurement. We note that, in the particular case illustrated in Figure 10.3, the probabilities of the two histories are 100% and zero respectively, but if the final polariser were differently oriented, these probabilities would change. However, if we arrange the detectors differently then we have to change the histories – i.e. use a different page in the book of maps. For example, suppose we insert a detector D_3 in the H channel between the two H/V polarisers. Assuming that D_3 allows the photon to pass through while recording its passage, there are now four consistent histories, corresponding to the four possible outcomes, as shown in the following table. The probability associated with each history is 25%.

	Before first polariser	Between the H/V polarisers, but before D_3	Between the H/V polarisers, but after D_3	Between the $\overline{\text{H/V}}$ and $\pm45°$ polarisers	After detection
1	$+45°,D_1. D_2,D_3$	$h,D_1. D_2 D_3$	$h,D_1. D_2 D_3*$	$+45°,D_1. D_2 D_3*$	$+45°,D_1*. D_2 D_3*$
2	$+45°,D_1. D_2,D_3$	$v,D_1. D_2 D_3$	$v,D_1. D_2 D_3$	$+45°,D_1. D_2 D_3$	$+45°,D_1*. D_2 D_3$
3	$+45°,D_1. D_2,D_3$	$h,D_1. D_2 D_3$	$h,D_1. D_2 D_3*$	$-45°,D_1. D_2 D_3*$	$-45°,D_1. D_2* D_3*$
4	$+45°,D_1. D_2,D_3$	$v,D_1. D_2 D_3$	$v,D_1. D_2 D_3$	$-45°,D_1. D_2 D_3$	$-45°,D_1. D_2* D_3$

As a final illustration, we consider making measurements on a pair of photons of the type discussed in Chapter 3. To recap on what we said there, pairs of photons are emitted from a source and have the property that their polarisations are always perpendicular. By this we mean that if we measure the H/V polarisation of the two photons, whenever one of them is found to be *h* the other will certainly turn out to be *v*. The question arose as to how one photon could 'know' what experiment we were performing on the other, and the Bell inequalities revealed that there are inevitable contradictions between the predictions of quantum physics and any model based on local hidden variables. Consider such an experiment, as illustrated in Figure 10.4, where H/V polarisation is measured on one photon, and the polarisation parallel or perpendicular to a direction at an angle ϕ to the horizontal is measured on the other. We have two particles and four detectors, so four consistent histories, as set out in Figure 10.4.

Referring back to the discussion in Chapter 3, the probabilities for these four histories are seen to be: (1) $\frac{1}{2}\cos^2\varphi$; (2) $\frac{1}{2}\sin^2\varphi$; (3) $\frac{1}{2}\sin^2\varphi$; and (4) $\frac{1}{2}\cos^2\varphi$, and we note that the sum of these four quantities is unity, as we expect if the family is to be consistent. If we were to change the value of ϕ then corresponding changes would occur in the histories, but we note again that this action would involve no physical change to the photon pair, only an informed choice of which map to use.

Thus, even when the measuring direction on one side is altered extremely rapidly, as in the Aspect experiment (see Figure 3.10), there is no physical communication between the photons: all that is required is that *we* change the map that we use so that it always contains the histories appropriate to

Fig. 10.4 Pairs of photons, whose polarisations are known to be perpendicular are emitted from a source located between two polarisers.
The consistent histories are:

	Before LH measurement	After LH, but before RH, measurement	After both measurements
1	$h, \phi_+, D_1, D_2, D_3, D_4$	$h, \phi_+, D_1*, D_2, D_3, D_4$	$h, \phi_+, D_1*, D_2, D_3*, D_4$
2	$h, \phi_-, D_1, D_2, D_3, D_4$	$h, \phi_-, D_1*, D_2, D_3, D_4$	$h, \phi_-, D_1*, D_2, D_3, D_4*$
3	$v, \phi_+, D_1, D_2, D_3, D_4$	$v, \phi_+, D_1, D_2*, D_3, D_4$	$v, \phi_+, D_1, D_2*, D_3*, D_4$
4	$v, \phi_-, D_1, D_2, D_3, D_4$	$v, \phi_-, D_1, D_2*, D_3, D_4$	$v, \phi_-, D_1, D_2*, D_3, D_4*$

the new set-up. We may recall Bohr's reply to the original EPR paper (see
Chapter 3):

... there is essentially the question of an influence on the very conditions that define
the possible types of prediction regarding the future behaviour of the system.

The examples that we have discussed all illustrate the essential features
of the consistent-histories approach to quantum measurement. Consistent
histories, particularly those associated with irreversible changes, are the possible representations of reality – i.e what may really happen. Even if there
is no measurement, we can still make (at least provisional) assignments of
reality to the histories in a consistent family.

Actualisation

Apart from the question of whether irreversible processes lead to perfectly
consistent histories, there may well be other grounds for criticising the
consistent-histories interpretation. To understand these, we reconsider some
of the ideas about the nature of scientific theory set out earlier in this chapter.
We drew a distinction between our model (map) of reality and reality itself:
the apple fell under gravity without having to do any mathematics! However,
in this example there is a clear one-to-one correspondence between each of
the parameters of our theory and an aspect of reality: the earth creates a
gravitational field that is 'really there' and the object responds to it. All the
information required for this action is out there in the physical world and our
calculations are simply our way of representing this. In contrast, there are a
number of possible histories in the quantum case; presumably one of them

corresponds to reality, but in the absence of measurement we cannot know which it is.

An important question is whether and how consistent histories resolve the measurement problem set out in Chapter 4. Consider again the archetypal measurement illustrated in Figure 10.2. If we treat the whole set-up as a quantum system, we may well predict that its final state will be a superposition of h, $D_1{}^*$, D_2 and v, D_1, $D_2{}^*$ rather than a collapse into one or other of these alternatives. Such a superposition is also a consistent history, so how has the measurement problem been resolved by the consistent-histories approach? The point is that the components of the superposition are also histories that are members of a consistent family. We use this family rather than the superposition because we are informed by our experience of the fact that the detectors always fire individually. The essential indeterminism of quantum physics means that there is often a choice of which consistent family to use and we choose the one that is appropriate for the experimental outcomes that we expect. The theory does not inevitably produce measurement outcomes, but we can always choose a representation in which these occur. The question still stands as to whether this is really an objective process or is this 'actualisation' of measurement outcomes put in 'by hand'? Some readers may well see a parallel with Einstein's comment on Bohr's position: that it was logically possible, but

... so very contrary to my scientific instinct that I cannot forego my search for a more complete conception.

For many years the leading critic of the Copenhagen interpretation was John Bell (of Bell's theorem fame – see Chapter 3). He emphasised that there was no dispute that this interpretation of quantum physics was correct 'for all practical purposes', but he strongly believed that science ought to be aiming for more than this. He entered a plea that our aim should be to describe the physical world as it is, not just the observations we are able to make on it. We should be trying to find a theory of 'beables', by which he meant parameters describing how the world could 'be', rather than just of 'observables', which describe how it can be 'observed'. The consistent-histories approach does not give any special status to observation as such and we might suggest that strong candidates for beables would be the component histories in a consistent family. John Bell died a few years after the consistent-histories interpretation had been developed, but there is no evidence that he paid it much attention. If he had, it is unlikely that this would have silenced his dissent, as he would surely have seen the need to choose an appropriate family as putting things in 'by hand'.

It is interesting to consider the relation between the consistent-histories approach and the many-worlds interpretation. A history in the former is equivalent to a branch in the latter. However, the corresponding ontologies are very different. In many worlds, instead of the branching points being the only contact between theory and reality, everything calculated for all possible branches has a counterpart in reality. If we assume that there is only one world, our map book contains a huge number of possible processes that are never actualised and there is no obvious place for this information to be physically stored. In contrast, many-worlds theory postulates a single map, large enough to contain all the information in the consistent-histories map book and with a one-to-one correspondence between all points on it and physical reality.

We saw in Chapter 6 that there are serious problems in understanding, or even defining, probability in the context of many-worlds theory. However, this is not a problem for the consistent-histories approach as each history in a given family is taken to have an associated calculable probability from the start. It is assumed that the randomness we observe in the quantum world is real and fundamental. The equations of quantum physics are fundamentally deterministic, in the sense that the future of any given history is fully determined. As we have seen, the family that we choose is determined by the experimental context and randomness comes in when one of the various histories within the family is actualised

If randomness and indeterminacy are intrinsic properties of the physical universe, this may be the best kind of theory we can hope for. Because if we were to introduce randomness at an earlier stage by modifying the quantum equations so that their solutions became very sensitive to the precise starting conditions, the outcomes might then appear to be random, but this would only be 'pseudo-randomness' – meaning that the same starting conditions inevitably lead to the same outcome provided we do the calculation accurately enough. As an analogy, when we use a computer to generate some apparently random numbers, we actually obtain the same set every time we run the program from the same starting point.[2] If, in contrast, quantum indeterminism is not pseudo-random, but a fundamental fact of nature, could it be represented in any other way than by a procedure at least similar to that discussed in this chapter? The consistent-histories approach claims that we have reached the point where a purely mathematical map is unable to give a unique description of the physical universe. It can, however, provide a map book containing all possible histories and their probabilities. Perhaps this is the best we can expect to achieve.

---

[2] To avoid this, some computers 'seed' their random number generator using the value of some voltage in the computer, which is fluctuating for essentially quantum reasons.

There seems to be little prospect of deciding experimentally between the different ontologies discussed so far, and we are largely free to choose the theory most suited to our philosophical prejudices. Those of the author are probably becoming fairly clear by now, but the last chapter of this book discusses them more explicitly.

11 · Illusion or Reality?

Now that we have completed our survey of the conceptual problems of quantum physics, what are we to make of it all? One thing that should be clear is that there is considerable scope for us all to have opinions and there is a disappointing lack of practicable experimental tests to confirm or falsify our ideas. Because of this, I intend to drop the use of the conventional scientific 'we' in this chapter and to use the first person singular pronoun wherever I am stating an opinion rather than describing an objective fact or a widely accepted scientific idea. This is not to say that everything in the earlier chapters has been free of personal bias, but I have tried to maintain a greater basis of objectivity there than will be appropriate from now on.

I first want to refer briefly to a way of thinking about philosophical problems known as 'positivism'. Encapsulated in Wittgenstein's phrase 'of what we cannot speak thereof should we be silent', positivism asserts that questions that are incapable of verification are 'non-questions', which it is meaningless to try to answer. Thus the famous, if apocryphal, debate between mediaeval scholars about how many angels can dance on the point of a pin has no content because angels can never be observed or measured so no direct test of any conclusion we might reach about them is ever possible. Opinions about such unobservable phenomena are therefore a matter of choice rather than logical necessity. Positivism can often exert a salutary beneficial influence on our thinking, cutting through the tangle of ideas and verbiage we (or at least I) sometimes get into, but if taken too far it can lead us to a position where many obviously acceptable, and apparently meaningful, statements are in danger of being dismissed as meaningless. An example is a reference to the past such as 'Julius Caesar visited Britain in 55 BC', or 'Florence Nightingale nursed the troops in the Crimean War'. There is no way such statements can be directly verified as we cannot go back in time to see them happening, but everybody who knows some history would believe them to be both true and significant and certainly not meaningless. Or what are we to make of a statement about the future such as 'the world will continue to exist after my death'? There is no way that I can directly verify this proposition, but I firmly believe it to be both meaningful and true. A positivist analysis can be useful, but it has to be employed with caution.

Other important concepts underlying any discussion of scientific theories are *simplicity* and *falsifiability*. The importance of simplicity (often known as 'Occam's razor') is a logical principle attributed to the mediaeval philosopher William of Occam. The principle states that no scientific model or theory should make more than the minimum necessary assumptions. Suppose that we make two observations of a moving object. If this is all the information we have then, in the absence of this principle, we could imagine anything we like about how the object moved between the two measurements. If, however, we apply Occam's razor, the simplest assumption is that the object has moved directly and smoothly between the two points. Often we have more information: for instance, if we know that the object was falling under gravity then our experience guides us to believe that the motion was actually one of uniform acceleration, as we assumed in drawing the graph in Figure 10.1, showing how the height of the falling apple changes in time. If we did an experiment that recorded this height at a few particular times only, we would apply Occam's razor to deduce that the calculated graph was also valid for those other times when we weren't looking, and reject (i.e. 'shave off') more complicated possible motions. For example, the object might have stopped in mid-air for a time and then accelerated rapidly, all while we were not watching it, and this behaviour would clearly be less simple than the standard Newtonian model. Clearly a test of simplicity is a necessary prerequisite of any scientific theory; however, it is not always infallible as it begs the question of how simplicity is to be defined and assessed. As we shall see, this underlies much of the ongoing controversy in the interpretation of quantum physics.

Turning to 'falsifiability', the importance of this concept was first stressed by the philosopher Sir Karl Popper (many years before he developed the ideas about the mind discussed in Chapter 4). He proposed that the difference between scientific and other kinds of statements is that it should always be possible to at least imagine an experimental test with a possible outcome that would show the statement to be false. Thus Newton's law of gravitation would be shown to be false if an apple or other object were to be released from rest, but not fall to the ground. He further suggested that every scientific theory must be considered provisional in the sense that we should believe it until it is falsified, and such falsification must always be a conceivable possibility. Even our strongly held belief that the sun will rise tomorrow morning rests on the assumption that the laws of physics will be the same then as they are today. We can conceive of the possibility that the sun may not rise and, if this were to happen, we should then develop a new theory, consistent with the observed facts, that would hold unless and until it was falsified in its turn. We see that falsification implies that we can imagine situations in which a theory might not hold, even if it always has done so far whenever we have actually tested it. Even if the apple always falls and the sun always rises, we can imagine that they might not. Statements of this type are known as 'counterfactuals' and were referred to briefly in Chapter 3. Counterfactual statements can be quite problematic philosophically, but in most circumstances we can understand what is meant.

The test of falsifiability can be used to distinguish scientific statements from what are known as 'metaphysical' statements made in other contexts (notably, perhaps, religion and politics) where it could be said that the tenets of the theory remain intact whatever facts are observed. Some would claim that unfalsifiable statements are meaningless, but Popper always argued against this and the positivist viewpoint in general. A form of positivism is clearly a very important part of the thinking behind the Copenhagen interpretation. If the very nature of the experimental set-up excludes any possibility of the measurement of a particular physical quantity, can that quantity have any 'reality' or in these circumstances is it just an 'illusion'? Certainly, as we saw in Chapter 2 and later in Chapter 10, the idea that a horizontally polarised photon is also polarised at $+45°$ or $-45°$ to the horizontal is a contradiction in terms – even in classical physics. Is it similarly meaningless to ask which slit a photon passed through in a two-slit interference experiment? As someone educated in (or perhaps brainwashed by) the Copenhagen tradition, I say 'yes, this is an illusion, the particle does not have a position – it is not really a particle – unless the experiment is designed to make a measurement of this property'. I am very aware that this kind of thinking does not come easily or naturally, but it seems to be forced on us by the development of quantum physics. On the other hand, in the nineteenth century some thinkers argued from a positivist viewpoint that the idea of matter being composed of atoms was a similarly meaningless postulate, which could not be directly tested. However, we all now accept the reality of the existence of atoms as a directly verifiable objective fact. Could it be that the Copenhagen interpretation is wrongly encouraging us to classify as illusion quantities that are perfectly real and will be observed when our knowledge and technology progress far enough? Thinking like this can make the idea of hidden variables seem both plausible and attractive – if it were not for the fact that no hidden-variable theory that preserves locality is capable of predicting the results of two-photon correlation experiments such as the Aspect experiment discussed in Chapter 3! Although the atomic postulate was not *necessary* to explain the known phenomena in the last century, it was always a perfectly tenable hypothesis. In contrast, Bell and Aspect and others have essentially falsified[1] the postulates of local hidden-variable theories. They are therefore untenable and cannot be believed in, however attractive such a belief might

---

[1] But note that some loophole may remain – see towards the end of Chapter 3.

be. If things had turned out differently and a successful hidden-variable theory had been developed on the basis of a simple model of the microscopic world, there is little doubt that it would have been generally accepted. The traditional quantum theory would very likely have been abandoned – even if both theories had made identical predictions of the results of all possible experiments. Positivists might have said that there was no meaningful distinction between the two approaches but, just as in the case of the atomic hypothesis, nearly everyone would have preferred a view based on a realistic model of the microscopic world. It is because this has not happened that I, along with most other physicists, have had to accept some at least of the Copenhagen ideas – not because we particularly wanted to but because this is the only way we can come near to describing the behaviour of the physical world. As Bohr pointed out a number of times it is nature itself rather than *our* nature that forces us into this new, and in many ways uncomfortable, way of thinking.

We see here an example of the application of Occam's razor: a local hidden-variable theory might have been considered simpler than the Copenhagen approach, but the Copenhagen approach is generally thought to be simpler than a non-local hidden-variable theory. However, a lot of subjective judgement is involved, and some hold out strongly against the consensus.

Nevertheless the Copenhagen interpretation leaves us with the measurement problem. If reality is only what is observed and if quantum physics is universal who or what does the observing? If the cat is a quantum object, what is it inside or outside the box that decides whether it is alive or dead or if it even exists? At this point the pure positivist might like us to go no further. 'It is meaningless to ask', she might say, 'whether it is your consciousness, the macroscopic apparatus or the irreversible change that causes the cat to be alive or dead – or indeed whether there is a multiplicity of cats each in its own universe – as there is no experiment to decide one way or the other'. Perhaps this is so, perhaps it is a non-question, but it is a fascinating non-question about which I have opinions that seem real to me at least.

We now turn to the question of consciousness, addressed in Chapter 5. Philosophers have long had difficulty proving that there is an objective real world 'out there' rather than that everything is just my sense impressions. However, the aim of science has always been to seek an objective description of the physical universe that can be consistently believed in if we so choose. To suggest that consciousness must fill an essential role in our understanding of the quantum world would run directly against this trend. It may be that a theory based on consciousness and subjectivism could be consistent with the observed facts, but I find its implications – such as the non-existence of a physical universe until a mind evolved (from what?) to observe it – quite unacceptable. I would prefer to believe almost any theory that preserved some form of objectivity.

What then about the many-worlds approach, where we assume that an unimaginably huge number of other universes 'really exists' and the idea that we have only one self in a single universe is an 'illusion'? When I first encountered this about 25 years ago, which was about 20 years after its invention, I shared most people's reaction that its ontological extravagance meant that it was not worthy of serious consideration. It is difficult to think of a more extreme example of the postulation of unobservable quantities to overcome a scientific problem. However, in common with many others, I have found myself taking it more seriously as the years went on. This is despite the fact that the postulate of a near infinite set of parallel universes seems, at first sight at least, a clear breach both Occam's razor and Popper's principle of falsifiability. Nevertheless, this seems to be the only approach that treats the equations of quantum physics as a single universal theory of the physical world, capable of describing all phenomena from the smallest to the largest without further postulates – 'cheap on assumptions' even if 'expensive on universes'. As we saw in Chapter 6, in a many-worlds theory we do not have to treat measurements, or irreversible changes in general, as different in kind from quantum events. Superpositions remain superpositions even when they interact with the measuring apparatus: all that happens is that branches result which are always unaware of each other. In the language of Chapter 10, instead of a map book we have a single map, although it has to be very large to contain all the information that would otherwise be in the book. However, because we have all these universes, we are postulating physical places in which all this information can be stored. Unless some experimental test can be devised, statements about the existence of parallel universes will remain unfalsifiable and therefore metaphysical in Popper's sense of the term, but not *ipso facto* meaningless unless we take a strong-positivist approach. Whether they should be acceptable in the light of Occam's razor is a judgement call. It is a matter of whether the principle should be applied to the ontology (the extravagant number of universes) or to the theoretical structure (with its economy of assumptions). In other words, before we can apply Occam's razor, we have to decide whether a single large map is simpler than a map book! The credibility of many-worlds theories has sometimes been damaged by misunderstandings that have not always been clarified by the efforts of its supporters. When Everett invented the theory, he stated that branching occurred following a 'measurement-like' event. This misled some of us to believe that branching has to be postulated, in much the same way as collapse is postulated in the Copenhagen interpretation. This was never Everett's intention, as has become clearer in the light of more recent developments in our understanding of decoherence, which show that branching is an inevitable part of the quantum physics of irreversible processes. It is the Copenhagen interpretation that contains the extra postulate of collapse into only one of these branches.

However, there is still the major problem of understanding the concept of probability in the many-worlds context. The whole language of probability refers to alternatives: either something is going to happen or it is not, but in many-worlds everything occurs somewhere. Current work in this area aims to show that it is consistent to define something that appears like a probability to an observer about to undergo a branching event. If a convincing resolution of this problem can emerge and if, as Everett believed, it can be coupled with a derivation of the magnitudes of these probabilities without additional postulates, then I should feel forced to take this approach very seriously.

One of the features of some writing on the quantum measurement problem is the statement, or at least implication, that there are only two possible views – subjectivism or many worlds – so that if one is rejected the other must be accepted. An example is an article written in *Psychology Today* by Harold Morowitz in 1980, where the subjective interpretation of the quantum measurement problem is described as the standard view of physicists, and this is contrasted with the mechanistic approach to the mind adopted by many modern biologists. In countering this, however, the writer Douglas Hofstadter[2] puts forward the many-worlds interpretation as the only alternative. If these were indeed the only alternatives then I should certainly choose the objective many-worlds rather than the subjectivist viewpoint, but I think we are a long way from having to make such a choice. The seeds of another approach were in fact already present in some of Niels Bohr's writing on the Copenhagen interpretation. Although his frequent references to 'the observer' have misled some writers into thinking that the Copenhagen interpretation is essentially subjective, this is not the case. Bohr was always at pains to emphasise the importance of the disposition of the measuring apparatus rather than any direct influence emanating from the experimenter. Thus, we have on the one hand

Every atomic phenomenon is closed in the sense that its observation is based on registrations obtained by means of suitable amplification devices with irreversible functions such as, for example, permanent marks on a photographic plate caused by the penetration of the electrons into the emulsion.

N. Bohr *Atomic Physics and Human Knowledge*, Wiley, New York (1958)

---

[2] In *The Mind's I*, cited in the bibliography for Chapter 5.

and

... it is certainly not possible for the observer to influence the events which may appear under the conditions he has arranged. To my mind, there is no alternative than to admit that, in this field of experience, we are dealing with individual phenomena and that our possibilities of handling the measuring instruments allow us only to make a choice between the different complementary types of phenomena we want to study.

N. Bohr, in *Albert Einstein: Philosopher-Scientist*, P. A. Schlipp (ed.), pp. 200–41, The Library of Living Philosophers, Evanston (1949)

On the other hand, Bohr also emphasised the importance of applying quantum physics to macroscopic objects, showing that this was necessary to preserve the consistency of quantum theory. As we saw in Chapter 7, if macroscopic objects were assumed to be completely classical then it would open the door to the possibility of making measurements on the atomic scale that are inconsistent with the uncertainty principle and the general laws of quantum physics. Nevertheless, Bohr does not appear to have seen the potential contradiction between these different approaches to the macroscopic world or to have seriously considered the problem of Schrödinger's cat.

Of course it doesn't really matter what Bohr or anyone else said or thought; science unlike much of law or theology defers not to past authority, but to the way nature is found to be. It should be clear from the later chapters in this book what a huge extrapolation is involved when we go from the measurement problem to either the subjective or the many-worlds postulate. Despite recent advances, our experimental knowledge of the application of quantum physics to the macroscopic world is very limited (see Chapter 7). It is important that the predictions of quantum physics continue to be tested in the macroscopic regime wherever possible. Unless and until there is experimental evidence for the breakdown of quantum physics in the macroscopic regime, I will continue to believe the likelihood of a falsification of quantum physics in this area to be small.

This brings us back to the idea of measurement as a thermo-dynamically irreversible process, discussed in Chapters 8, 9 and 10. The qualitative difference between, on the one hand, the quantum process in which a photon or electron passes through a polariser or interference slits and, on the other hand, the measurement process recording of the arrival of the particle on the emulsion of a photographic plate is enormous. This view of measurement may be open to criticism on the basis that an inevitable, if eventual, Poincaré recurrence requires quantum coherence to be maintained, so that it is never correct to say that a measurement has finally been made. However, as we saw in Chapter 9, this argument may well not apply if the system is chaotic rather than ergodic. There is certainly a practical operational distinction between thermodynamically irreversible changes and the reversible changes undergone by systems composed of a few particles – or indeed by composite many-particle systems moving coherently as in SQUIDs. If we pursue the view that we should learn our ways of thinking about natural phenomena from the way that nature behaves, then surely we should take very seriously the proposition that this distinction is a fundamental property of the physical world and not one we make for our own convenience. It is because of this that I have found Prigogine's ideas attractive. Instead of starting from the microscopic abstraction and trying to derive laws from it that describe both the coherent evolution of quantum states and the changes associated with measurements, he suggested that we do the opposite. Why not try taking as the primary reality those processes in the physical world that are actually observed – the cat's death, the blackening photographic emulsion, the formation of a bubble in the liquid-hydrogen bubble chamber? We may then treat as 'illusion', or at least as an approximation to reality, sub-atomic processes such as a photon passing through both slits or an elementary particle changing by two quantum processes simultaneously. Of course, as was discussed in Chapter 9, this would imply a revolutionary change in our thinking: the fundamental reality would not now be the existence of the physical world but the irreversible changes occurring in it – not 'being' but 'becoming'. Reversible 'events' that by definition leave no permanent records are then illusions – not just to us but to the whole universe, which undergoes no irreversible change as a result of their 'occurrence'.

The more recent work on consistent histories and decoherence, discussed in Chapter 10, in a sense closes the circle. Although developed as a fresh interpretation of quantum physics, the outcome has been a synthesis of the Copenhagen interpretation with the role of irreversibility. I find this approach greatly illuminates what Bohr probably intended. 'The very conditions that define the possible types of prediction regarding the future behaviour of the system' means that we must construct our theoretical model, or map, in terms of appropriate consistent histories. An essential part of this approach is that it treats indeterminism and the probabilities of irreversible changes as the main contact point between quantum theory (the map book) and the physical world.

One reason I find this way of looking at things attractive is the importance it gives to a traditional, serial way of considering the nature of time. Since Einstein it has become fashionable to look at time as just another dimension and to talk about 'space–time'. However, time and space are not equivalent concepts, even in the theory of relativity, and although we could imagine a world with a greater (or lesser) number of spatial dimensions than three, it is very hard to imagine a world with no, or more than one, time dimension. Without the possibility of change the idea of existence is arguably meaningless, so for me at least there is no being without becoming. If, as a result of the modern work on irreversible processes, we were to be led to a fundamental physics that took as its central theme the idea that time really does flow in one direction, I at least should certainly welcome it.

I also like this approach to nature because it is fundamentally a scientific approach. Since the development of quantum theory, too many people, some of whom should have known better, have used it to open a door to some form of mysticism. The outrageous leap from the measurement problem to the necessity for the existence of the human soul is just one extreme example of this. Another, as we saw at the end of Chapter 4, is the suggestion that the delocalisation of quantum states can be used to 'explain' extra-sensory perception and all sorts of other 'paranormal' phenomena. Much of this is based on a misunderstanding of quantum physics, but just about all of it is dispelled if the irreversible measuring processes are taken as the primary reality, because, although often indeterministic, these processes are objectively real and are subject to the principle of locality and the theory of relativity. Although we are a long way from any detailed understanding of the workings of the human brain, the underlying processes are clearly complex and chaotic and have more in common with measurement-type irreversible changes than with coherent quantum phenomena. It seems to me, therefore, that attempts to explain mental operations in traditional quantum terms are doomed to failure and that the idea sometimes put forward of a parallelism between quantum physics and psychology is only a superficial analogy at best.

If you have persisted to this point, it should be abundantly clear that there is no consensus answer to the problems of quantum physics. This fact is emphasised by the nature of the literature in this area, much of which is remarkably polemical in tone. Some stoutly defend the idea that realism requires hidden variables, while others believe that the case for many worlds is overwhelming. There is a disappointing lack of dialogue: many writers defend their particular viewpoint by stressing its advantages and avoiding addressing the quite valid criticisms made of it. Despite my own views summarised in this chapter, I believe that it is still too early to be dogmatic, because we are just beginning to understand the quantum behaviour of our chaotic universe. My hope would be that further study will open up new possibilities of experimental testing and that the distinction between what is illusion and what is reality will continue to be pursued by scientists as well as philosophers.

Further reading

This is not intended to be an exhaustive bibliography of the subject, but a general guide to some of the large number of relevant publications. All the references contained in the first edition are included below, though not all are still in print. The later references form a somewhat arbitrary selection from what is now a large corpus.

General bibliography

D. Bohm *Causality and Chance in Modern Physics* (Routledge and Kegan Paul, London, 1959 and 1984). This volume discusses the problems of quantum theory as they appeared fifty years ago to the leading proponent of hidden-variable theories.

P. C. W. Davies *Other Worlds* (Dent, London, 1980) discusses quantum ideas along with other developments in modern physics. Written by a professional physicist for the general reader.

B. d'Espagnat *Conceptual Foundations of Quantum Mechanics* (Benjamin, Massachusetts, 1976). This book and its author played a large part in a revival of interest in the conceptual problems. Written at a level suitable for the professional physicist competent in mathematics.

B. d'Espagnat *In Search of Reality* (Springer, New York, 1983). In this volume the author explains his ideas in a non-mathematical way.

A. I. M. Rae *Quantum Mechanics* (5th edn., Taylor & Francis, 2008). This university textbook by the present author concentrates mainly on principles and applications but also includes a chapter on the conceptual problems.

J. A. Wheeler and W. H. Zurek (eds.) *Quantum Theory and Measurement* (Princetown University Press, Princeton, 1983). A collection of many original articles published between 1926 and 1981. Of particular interest are the extracts from the Bohr–Einstein dialogue and the extensive bibliography.

J. Al-Khalili *Quantum: A Guide for the Perplexed* (Weidenfeld and Nicholson, London, 2003). A modern, lavishly illustrated, popular account of the principles and applications of quantum physics.

J. Baggott *Beyond Measure: Modern Physics, Philosophy and the Meaning of Quantum Theory* (Oxford University Press, Oxford, 2003). This textbook is intended for physics and philosophy undergraduates and reviews the whole field.

Specific bibliography

1 · Quantum physics

J. Powers *Philosophy and the New Physics* (Methuen, London, 1982). A readable discussion of the development of scientific and philosophical ideas over the last few centuries.

T. Hey and P. Walters *The Quantum Universe* (Cambridge University Press, Cambridge, 1987). An excellent account of the successful application of quantum physics to a wide range of physical situations.

2 · Which way are the photons pointing?

A. P. French and E. F. Taylor *An Introduction to Quantum Physics* (Nelson, Middlesex, 1978). This university textbook includes a discussion of the polarization properties of photons, and of polarized light in general, that goes a bit further than the present volume.

3 · What can be hidden in a pair of photons?

J. F. Clauser and A. Shimony 'Bell's Theorem: experimental tests and implications' (*Reports on Progress in Physics*, **41**, 1881–1927, 1978). This review article contains a thorough discussion of the various proofs of Bell's theorem and describes the various experiments performed up to that time, which was before the Aspect experiments.

B. d'Espagnat 'The quantum theory and reality' (*Scientific American*, **241** (11), 128–66, 1979). This article gives a semi-popular account of Bell's theorem and its implications for non-separability. Written before the Aspect experiments.

5 · Is it all in the mind?

K. Popper and J. C. Eccles *The Self and its Brain* (Springer, Berlin, 1977). Sets out the ideas discussed in the first part of the chapter.

D. R. Hofstadter and D. C. Dennett (eds.) *The Mind's I* (Penguin, Middlesex, 1981). This collection of essays by a number of writers on the problem of consciousness contains a commentary by the authors arguing for a model of the mind based on the ideas of artificial intelligence.

J. Searle *Minds, Brains and Science* (BBC, London, 1984). The 1984 Reith lectures, which argue against the separation of mind and brain but oppose the idea of a computer-like model for the mind.

W. Sargeant and H. J. Eysenck *Explaining the Unexplained* (Weidenfeld and Nicholson, London, 1982). This survey of paranormal phenomena includes a chapter describing the attempts to explain them by quantum physics.

M. Gardner *Science: Good, Bad and Bogus* (Oxford University Press, Oxford, 1983). Although it contains few references to quantum physics, this collection of Martin Gardner's writings over the years contains a number of robust criticisms of research into paranormal phenomena and is a refreshing antidote to the book by Sargeant and Eysenck cited above.

R. Penrose *The Emperor's New Mind* (Oxford University Press, Oxford, 1989) and *Shadows of the Mind* (Oxford University Press, Oxford, 1994) develops his view that the workings of consciousness cannot be simulated by a computer and considers the possible relevance of quantum physics to this.

6 · Many worlds

B. S. DeWitt and N. Graham (eds.) *The Many-Worlds Interpretation of Quantum Mechanics* (Princeton University Press, Princeton, 1973). A collection of a number of the original papers on this topic, including Everett's Ph.D. thesis.

D. Deutsch *The Fabric of Reality* (Penguin, Middlesex, 1997). A wholehearted supporter of many worlds explains his ideas on quantum physics, among other things, but without any serious attempt to answer criticisms.

7 · Is it a matter of size?

A. J. Leggett Schrödinger's cat and her laboratory cousins (*Contemporary Physics*, **25**, 583–98, 1984). This review article describes Professor Leggett's idea of a possible macroscopic resolution of the measurement problem and how it may be tested by experiments on SQUIDs.

A. J. Leggett 'Testing the limits of quantum mechanics: motivation, state of play, prospects' (*Journal of Physics: Condensed Matter* **14**, R415–R451, 2002). An excellent review of the state of play in theory and experiment.

8 · Backwards and forwards

P. C. W. Davies *Space and Time in the Modern Universe* (Cambridge University Press, Cambridge, 1977). This wide-ranging semi-popular account of modern physics includes an excellent discussion of irreversibility and time asymmetry.

P. T. Landsberg (ed.) *The Enigma of Time* (Hilger, Bristol, 1982). This collection of papers spans many aspects of the nature of time, including irreversibility.

P. C. W. Davies *About Time* (Viking, London, 1995). A popular exposition.

9 · Only one way forward?

I. Prigogine *From Being to Becoming* (Freeman, San Francisco, 1980). This book forms much of the basis of the discussion in the chapter as well as describing a number of other interesting manifestations of irreversible thermodynamics.

I. Prigogine and I. Stengers *Order out of Chaos* (Heinemann, An, 1984). A popular version of the book cited above.

10 · Can we be consistent?

R. B. Griffiths *Consistent Quantum Theory* (Cambridge University Press, Cambridge, 2002). The proponent of 'consistent histories' explains his ideas.

R. Omnes *The Interpretation of Quantum Mechanics* (Princeton University Press, Princeton, 1994) and *Understanding Quantum Mechanics* (Princeton University Press, Princeton, 1999). Exhaustive and quite technical accounts of the consistent-histories approach.

Index

Printed in the United States
by Baker & Taylor Publisher Services